Linear IC Applications

A Designer's Handbook

Linear IC Applications

A Designer's Handbook

Joseph J. Carr

ELSEVIER SCIENCE Ltd
The Boulevard, Langford Lane
Kidlington, Oxford OX5 1.GB, UK

© 2006 Elsevier Science Ltd. All rights reserved.

This work is protected under copyright by Elsevier Science, and the following terms and conditions apply to its use:

Photocopying
Single photocopies of single chapters may be made for personal use as allowed by national copyright laws. Permission of the Publisher and payment of a fee is required for all other photocopying, including multiple or systematic copying, copying for advertising or promotional purposes, resale, and all forms of document delivery. Special rates are available for educational institutions that wish to make photocopies for non-profit educational classroom use.

Permissions may be sought directly from Elsevier Science Global Rights Department, PO Box 800, Oxford OX-5 I DX, UK; p hone: (+ 44) 1.865 843830, fax: (+ 44) 1.865 853333, e-mail permissions @ e I sevier. co. uk. You may also contact Global Rights directly through Elsevier's home page (http://www.elsevier.com), by selectIng 'Obtaining PermIssions'.

In the USA., users may clear permissions and make payments through the Copyright Clearance Center, Inc., 222 Rosewood Drive, Danvers, MA. 01923, USA; phone: (+ 1) (978) 7508400, fax: (978) 7504744, and In the UK through the Copyright Licensing Agency Rapid Clearance Service (CI.ARCS), 90 Tottenham Court Road, London W1P OLP, UK; phone: (+44) 207 631. 5555; fax: (+44) 207 6315500. Other countries may have a local reprographic. rights agency for payments.

Electronic Storage or Usage
Permission of the Publisher is required to store or use electronically any material contained in this work, including any chapter or part of a chapter.

Except as outlined above, no part of this work may be reproduced, stored in a retrieval system or transmitted in any form or by any means, electronic, mechanical, photocopying, recording or otherwise, without prior written. permission of the Publisher. Address permissions requests to: Elsevier Science Global Rights Department, at the mail, fax and e-mail addresses noted above.

First published in 1996 by Newnes.
☐Library of Congress Cataloging in Publication Data .A catalog record from the Librar of Congress has been applied for

☐ISBN: 0-7506-3370-

☐requirements of A NSI/NISO Z39.48-1.992 (Permanence of Paper)

☐Printed in The Netherlands

Contents

Preface

Free supplemental software offer

1. Nonlinear (diode) applications of linear IC devices 1
2. Signal processing circuits 32
3. Measurement and instrumentation circuits 67
4. Integrated circuit timers 114
5. IC data converter circuits and their application 148
6. Audio applications of linear IC devices 175
7. Communications applications of linear IC devices 211
8. Analog multipliers and dividers 254
9. Active filter circuits 281
10. Troubleshooting discrete and IC solid-state circuits 323

Index 345

Preface

From time-to-time, one hears that analog electronics, i.e. the domain of linear IC devices, is dead, and that digital electronics is taking over every task. While it is true that digital electronics is growing rapidly, and has already taken over many functions previously performed in analog circuits, that doesn't mean that analog electronics is ready to die. A spokesman for a major supplier of special-purpose linear IC devices told me that the company's analog product lines are growing every year at a rate comparable to the digital product lines in both sales volume and monetary terms.

An argument in favor of analog electronics is that there are still jobs that are either best done in analog circuits, or are more cost effective when done in analog circuits rather than computers. For example, the human electrocardiograph (ECG) machine used by physicians must be able to strip off a one millivolt (1 mV) physiological signal riding on a 1,000 to 2,000 mV DC component formed by the half-cell potential of the metallic electrode against an electrolytic human skin. Indeed, the variation in the half-cell potential is greater than the signal potential. Even computerized ECG monitors often use an AC-coupled analog amplifier ahead of the analog-to-digital converter (ADC) in order to boost the signal and strip off the DC component. One designer pointed out that a relatively cheap 8-bit ADC could be used when the ECG signal was amplified to about 1,000 mV, and a 10-bit ADC would yield better results than required. But if the signal was not amplified and stripped of its DC component, then a 24-bit ADC would be needed. The cost ratio in manufacturing lot quantities was about 8:1 between the analog solution and the straight digital solution.

A second example is in the processing of phototransistor and photodiode signals. Typically, such sensors produce nanoampere signal levels riding on a DC pedestal on the order of milliamperes; 100:1 and 1,000:1 ratios are not uncommon. In these circuits, it is usually best to use a linear analog circuit to process the signal prior to applying to the digital circuit. Indeed, many digital instruments require a relatively extensive analog subsystem in order to work properly. Perhaps that's why the IC manufacturer spokesman was optimistic for the future.

This book is about practical applications of linear IC circuits. Although most of the circuits are based on the ubiquitous operational amplifier, other devices are examined as well. The material in this book will allow you to design circuits for the applications covered. But more than that, the principles of design for each class of circuit are transferable to other projects that are similar in function, if not in detail.

Joseph J. Carr, B.Sc., MSEE

Free supplemental software offer

Accompanying this book is a suite of supplemental software written by the author to assist in the design of linear IC circuits. A collection of circuits are offered covering a range of applications. For each circuit a simple slider bar operation allows the user to alter the parameters until a desired combination is achieved. Topics covered include inverting and noninverting amplifiers, differential amplifiers, instrumentation amplifiers, waveform generators and filters. This software is available for downloading from Butterworth-Heinemann's website:

http://www.butterworth.heinemann.co.uk/carr/carr.html

It is also available on disk for a small handling charge. Please contact Duncan Enright, Newnes Publisher, at Butterworth-Heinemann:

e-mail: duncan.enright@bhein.rel.co.uk
(make the subject of your message 'Carr Software Offer')

or write to:

Duncan Enright, Newnes Publisher
Butterworth-Heinemann
Linacre House, Jordan Hill
Oxford OX2 8DP, UK.

CHAPTER 1

Nonlinear (diode) applications of linear IC devices

OBJECTIVES

1. Learn the function of, and applications for, the precise diode circuit.
2. Understand how zero-bound and dead-band circuits work.
3. Be able to describe the operation of active clipper and clamper circuits.
4. Understand the peak follower and sample and hold circuits.

1.1 PRE-QUIZ

These questions test your prior knowledge of the material in this chapter. Try answering them before you read the chapter. Look for the answers (especially those you answered incorrectly) as you read the text. After you have finished studying the chapter try answering these questions again, and those at the end of the chapter (see Section 1-10).

1. Draw the V_o versus V_{in} curves for both ideal and practical rectifiers.
2. Give two reasons why an active 'precise rectifier' might be preferred over a single diode rectifier.
3. Design a zero-bound amplifier that has a threshold point of +3 volts.
4. Draw the schematic for a peak follower circuit.

1.2 INTRODUCTION

Although the operational amplifier is well known for its capabilities as a linear amplifier, there are also numerous nonlinear applications for which the op-amp is also well suited. In this chapter you will study circuit techniques that are commonly employed in fields as diverse as instrumentation, control, and communications. Of particular interest here are circuits in which PN junction diodes are used: *precise rectifiers, bounded circuits, clippers/clampers,* etc.

1.3 REVIEW OF THE PN JUNCTION DIODE

The PN junction diode is the oldest solid-state electronic component available. Indeed, naturally occurring 'diodes' of *galena* crystals (lead sulphide, or PbS) were used prior to World War I as the demodulator (or 'detector') in crystal set radio receivers. During World War II radar research led to the development of the 1N34 and 1N60 germanium video detector diodes, and the 1N21 and 1N23 microwave diodes.

The PN junction diode ideally has a transfer characteristic like Fig. 1-1A. When the anode is positive with respect to the cathode (Fig. 1-1B), the diode is forward biased so conducts current. Alternatively, when the anode is negative with respect to the cathode (Fig. 1-1C), the diode is reverse biased and no current flows. Figure 1-1D shows the effect of this unidirectional current flow on a sinewave input signal. Notice in the *halfwave rectified* output only the positive peaks are present.

Real diodes fail to meet the ideal in several important respects. Figure 1-2 shows a transfer characteristic for a practical, non-ideal diode. For the ideal diode the reverse current is always zero, while in real diodes there is a minute leakage current (I_L) flowing backwards across the junction. A manifestation

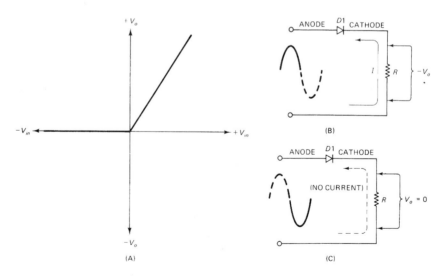

FIGURE 1-1 (A) V_o versus V_{in} curve for ideal rectifier; (B) current flows on positive halfcycle; (C) no current flows on negative halfcycle; (D) input and output waveforms of halfwave rectifier.

2 NONLINEAR (DIODE) APPLICATIONS OF LINEAR IC DEVICES

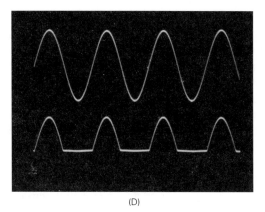

(D)

FIGURE 1-1 (continued)

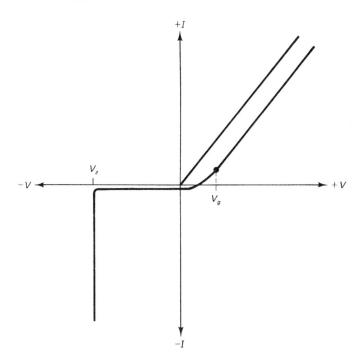

FIGURE 1-2 Ideal and real rectifier transfer function.

of this current can be seen by measuring the forward and reverse resistances of a PN junction diode. The forward resistance is very low, while the reverse resistance is very high... but not 'infinite' as one might expect from a supposed open circuit.

Another departure from the ideal in the reverse bias region is the avalanche point (V_z) at which reverse current flow increases sharply. At this point, the reverse bias voltage is too great and causes breakthrough. When carefully regulated, the breakdown potential is both sharply defined and reasonably stable except for a slight temperature dependence. In such cases the device is called a *zener diode*, and is used as a *voltage regulator*.

In the forward biased region there are other anomalies that depart from the ideal. In the ideal case there is an ohmic relationship between current flow and applied forward voltage. Similarly, there is a linear relationship between applied forward voltage (V_f) and output voltage V_o. In real diodes, however, there is a significant departure from the ideal transfer characteristic. Between zero volts and a critical junction potential (V_g) the characteristic curves are quite nonlinear. The actual value of this potential is a function of both the type of semiconductor material used and the junction temperature. In general, V_g will be 0.2 to 0.3 volts for germanium (Ge) diodes (1N34, 1N60, etc.), and 0.6 to 0.7 volts for silicon (Si) diodes (1N400x, 1N914, 1N4148, etc.). In the 0 to V_g region the diode forward resistance is a variable function of V_f and T, and the I versus V_f characteristic is logarithmic. Above V_g the characteristic becomes more nearly linear.

1.4 PRECISE DIODE CIRCUITRY

A *precise diode circuit* combines an active device such as an operational amplifier with a pair of diodes to essentially 'servo-out' the errors of the non-ideal diode. Two advantages obtain from this arrangement. First, the circuit will rectify very small AC signals between zero volts and V_g (i.e. $0 < V < V_g$, about 0.65 volts). Second, the rectification will be more nearly linear than is the case with just the diode alone, even in the diode's ohmic range.

Figure 1-3 shows the circuit for the inverting halfwave precise rectifier. A basic assumption in this circuit is that load impedance R_L is purely resistive, and therefore contains no energy production or storage elements. The circuit is essentially an inverting follower amplifier with two PN junction diodes ($D1$ and $D2$) added. Halfwave rectification occurs because the circuit offers two different gains that are dependent on the polarity of the input signal. For positive values of V_{in} the gain (V_o/V_{in}) is zero, while for negative values of V_{in} the voltage gain is R_f/R_{in}.

Consider operation of the circuit for positive values of V_{in}. The noninverting input (+IN) is grounded, so is at zero volts. By the properties of the ideal op-amp we must consider the inverting input (−IN) as if it is also grounded ($V_a = 0$). Recall that this concept is called a *virtual ground*. Thus, differential voltage V_d is zero.

When $V_{in} > 0$, i.e. when it is positive, current $I1 = +V_{in}/R_{in}$. In order to maintain the equality $I1 + I3 = 0$, conserving Kirchhoff's current law (KCL), the op-amp output voltage V_b swings negative, but is limited by the $D1$ junction potential to V_g (about 0.6 to 0.7 volts). With $V_b < 0$, even by only 0.6 to 0.7 volts, diode $D2$ is reverse biased and therefore cannot conduct.

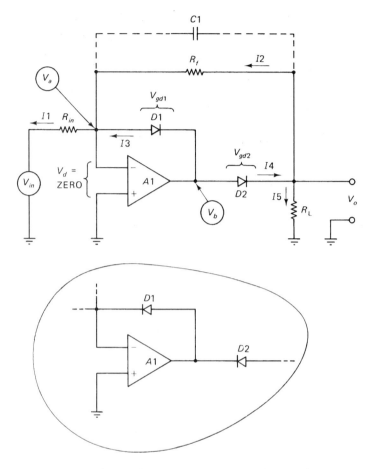

FIGURE 1-3 Precise rectifier circuit (inset shows reverse polarity version).

Currents $I2$, $I4$ and $I5$ are zero. Thus, for positive values of V_{in} the output voltage V_o is zero.

Now consider operation for $V_{in} < 0$. Under this input condition op-amp output voltage V_b swings positive, forcing diode $D1$ to become reverse biased, and $D2$ to conduct. In order to conserve KCL (as before) for this case $(I1 + I2 = 0)$, current $I2$ will have the same magnitude but opposite direction relative to $I1$. Because $V_{in}/R_{in} = -V_o/R_f$, the voltage gain $(A_v = V_o/V_{in})$ reduces to $-R_f/R_{in}$, as is appropriate for an inverting amplifier. Thus, the gain for negative input voltages $(V_{in} < 0)$ is $-R_f/R_{in}$, while for positive input voltage $(V_{in} > 0)$ it is zero. From that difference comes halfwave rectification.

The voltage drop across diode $D2$ is about $+0.6$ to $+0.7$ volts, and is 'servoed-out' by the fact that $D2$ is in the negative feedback loop around

A1. Voltage V_b is correspondingly higher than V_o in order to null the effects of V_{gd2}.

The precise rectifier is capable of halfwave rectifying very low-level input signals. The minimum signal allowed is given by:

$$V_{in} > \frac{V_g}{A_{vol}} \tag{1-1}$$

where:

V_{in} is the input signal voltage (V)
V_g is the diode junction potential (0.6 to 0.7 volts)
A_{vol} is the open-loop gain of the amplifier.

In Eq. (1-1) the term A_{vol} is the open-loop gain, which for DC and low-frequency AC signals is extremely high. But at some of the frequencies at which precise diodes operate, the input frequency is a substantial fraction of the gain–bandwidth product, so A_{vol} will be less than might otherwise be true. For example, if the gain–bandwidth product is 1.2 MHz, the gain at 100 Hz is 12 000. But at 1000 Hz, a typical frequency for precise rectifier operation, the gain is only 1200.

EXAMPLE 1-1

Calculate the minimum signal that will be accurately rectified in a precise rectifier circuit in which the open-loop gain is 3000. Assume that silicon diodes are used, so $V_g = 0.7$ volts.

Solution

$$V_{in} > \frac{V_g}{A_{vol}}$$

$$V_{in} > \frac{0.7 \text{ volts}}{3000} = 0.00023 \text{ volts} \qquad \blacksquare$$

Circuit operation of the precise rectifier is shown by the waveforms in Fig. 1-4. If a sinewave is applied (Fig. 1-4A), the output voltage V_o will be zero from time $T1$ to $T2$ (positive input voltage), while V_b will rest at $-V_g$ (about -0.6 to -0.7 volts). Between $T2$ and $T3$ the input is negative, so V_o will be a positive voltage with a halfwave sine shape (Fig. 1-4B). But note the behavior of V_b, the op-amp output (Fig. 1-4C). From $T1$ to $T2$ the output rests at $-V_g$, but at $T2$ it snaps to a value of $2 \times V_g$ to the positive. The halfwave sine shape rests on top of the $+V_g$ offset caused by V_{gd2}. Figure 1-4D shows this same situation in the form of the transfer characteristic (V_o versus V_{in}).

The circuit of Fig. 1-3 as shown will rectify and invert negative peaks of the input signal. In order to accommodate the positive peaks one need only reverse the polarity of diodes $D1$ and $D2$.

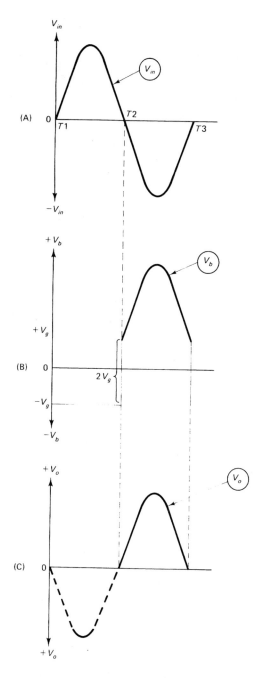

FIGURE 1-4 (A) input voltage waveform; (B) halfwave sine shape; (C) only positive halfcycle is output; (D) input and output waveforms; (E) V_o versus V_{in}; (F) halfwave rectified outputs for both polarities.

NONLINEAR (DIODE) APPLICATIONS OF LINEAR IC DEVICES

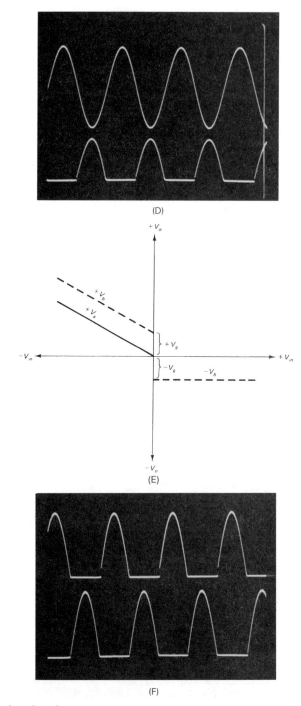

FIGURE 1-4 (*continued*)

8   NONLINEAR (DIODE) APPLICATIONS OF LINEAR IC DEVICES

1-4.1 Precise diodes as AM demodulators

One of several applications of the precise diode circuit is demodulation or detection of amplitude modulated (AM) carrier signals. An AM signal (Fig. 1-5) is one in which a sinusoidal carrier signal of frequency F_c is varied in amplitude by a lower frequency modulating frequency F_m. The modulating signal might be either sinusoidal or non-sinusoidal.

The most familiar use of AM is in radio broadcasting and communications. Not so familiar, perhaps, is the use of AM in non-radio applications such as instrumentation. One common use is in certain AC-excited Wheatstone bridge transducers. In one popular pressure transducer the excitation signal is a 10 volt peak-to-peak, 400 Hz sinewave. The output of the transducer is an AC signal that is proportional to the excitation voltage and the applied pressure. Thus, the pressure signal is used to amplitude modulate the 400 Hz carrier signal.

Demodulation of an AM signal is usually done by *envelope detection*. This type of detector is basically a halfwave rectifier and a low-pass filter. The precise diode circuit of Fig. 1-5 can be used to demodulate AM signals, and offers an advantage because of its ability to accommodate weak signals. Low-pass filtering is obtained by shunting capacitor $C1$ across feedback resistor R_f.

For any given type of operational amplifier there is a maximum carrier frequency that can be accepted for AM demodulation. This frequency is related to the gain–bandwidth product of the particular device selected:

$$F_{c(max)} = \frac{F_t}{100|(A_v)|} \qquad (1\text{-}2)$$

where:

$F_{c(max)}$ is the maximum allowable carrier frequency

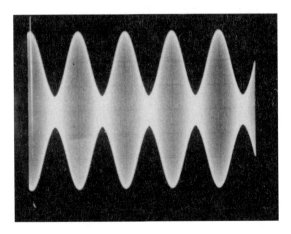

FIGURE 1-5 Amplitude modulated carrier waveform.

F_t is the device gain-bandwidth product (i.e. the frequency at which $A_{vol} = 1$)
$|A_v|$ is the absolute value of the closed-loop voltage gain

EXAMPLE 1-2

A BiMOS operational amplifier with a gain-bandwidth product of 12 MHz is used as a precise rectifier AM demodulator. Calculate the maximum carrier frequency that the circuit can handle in a gain-of-100 circuit.

Solution

$$F_{c(max)} = \frac{F_t}{100 \text{ abs}(A_v)}$$

$$F_{c(max)} = \frac{1.2 \times 10^7 \text{ Hz}}{(100)(\text{abs}(-100))}$$

$$F_{c(max)} = \frac{1.2 \times 10^7}{(100)(100)} = 1200 \text{ Hz} \quad\blacksquare$$

The operational amplifier slew rate (S) must be sufficient to handle the input signal carrier frequency. In general:

$$S > \frac{\Delta V_{in}}{\Delta t} \quad (1\text{-}3)$$

or, for the specific case of sinusoidal carriers:

$$S > 2\pi F_c |(A_v)| V_{in(peak)} \quad (1\text{-}4)$$

where:
S is the slew rate in volts per microsecond (V/μs)
ΔV_{in} is the change in input voltage over time Δt
Δt is the time over which V_{in} changes

The value of capacitor $C1$ (Fig. 1-3) is found from:

$$C1 = \frac{\sqrt{F_c/F_m}}{2\pi F_c R_f} \quad (1\text{-}5)$$

EXAMPLE 1-3

A 1000 Hz sinusoidal carrier is modulated by a 15 Hz sinewave. Calculate $C1$ when 10 kohm resistors are used for both R_f and R_{in}. Assume a peak signal voltage of 2 volts.

Solution
(a)
$$S > 2\pi F_c|(A_v)|V_{in(peak)}$$
$$S > (2)(3.14)(100 \text{ Hz})(10 \text{ k}\Omega/10 \text{ k}\Omega)(2 \text{ volts})$$
$$S > 12\,560 \text{ V/s} > 0.013 \text{ V/μs}$$

(b)
$$C1 = \frac{\sqrt{F_c/F_m}}{2\pi F_c R_f}$$

$$C1 = \frac{\sqrt{1000 \text{ Hz}/15 \text{ Hz}}}{(2)(3.14)(1000 \text{ Hz})(10\,000 \text{ }\Omega)}$$

$$C1 = \frac{\sqrt{66.7}}{6.3 \times 10^7} = 0.013 \text{ }\mu\text{F} \qquad \blacksquare$$

1-4.2 Polarity discriminator circuits

A *polarity discriminator* is a circuit that will produce outputs that indicate whether the input voltage is zero, positive or negative. Applications for this type of circuit include alarms, controls and instrumentation.

Figure 1-6A shows a typical polarity discriminator circuit. The basic configuration is the inverting follower op-amp circuit, but with two negative feedback circuits. Each feedback path contains a diode, but since the diodes are connected in the opposite polarity sense, the polarity of the output potential will determine which one conducts and which is reverse biased. Consider first the case where the input signal V_{in} is positive. In this case, current I_{in} flows away from the summing junction toward the source, and has a magnitude of $+V_{in}/R_{in}$. The output terminal of the op-amp will swing negative, causing diode $D1$ to be reverse biased, and $D2$ to be forward biased.

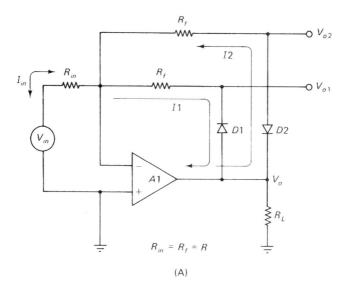

FIGURE 1-6 (A) polarity discriminator circuit; (B) circuit waveforms; (C) oscilloscope photo of circuit waveforms.

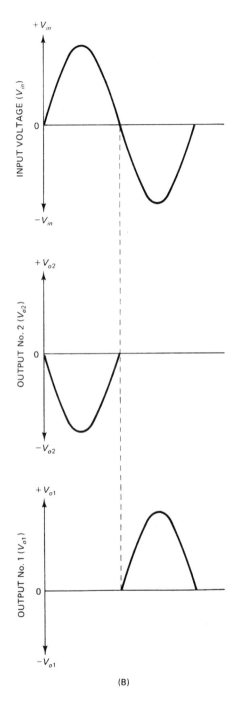

FIGURE 1-6 (*continued*)

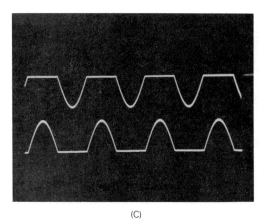

(C)

FIGURE 1-6 (*continued*)

Current $I1$ is zero, and $I2$ is equal to V_{o2}/R_f. Output voltage V_{o2} is negative, and has a value $V_{o2} = V_o - 0.6$ volts; output voltage V_{o1} is zero.

Now consider the opposite case, i.e. where V_{in} is negative. The current flows away from the source toward the summing junction. The output terminal of the op-amp swings positive, causing diode $D1$ to become forward biased, while $D2$ is reverse biased. Current $I2$ is there zero, while $I1$ is V_{o1}/R_f. In this case V_{o1} is positive, while V_o is zero.

The operation of this circuit can be seen in the waveforms shown in Fig. 1-6B, and in the transfer characteristics shown in Fig. 1-6C.

1-4.3 Fullwave precise rectifier

The *fullwave rectifier* uses both halves of the input sinewave. Recall that the halfwave rectifier removes one polarity of the sinewave; the fullwave rectifier preserves it. Figure 1-7 shows the relationships present in a fullwave rectifier circuit. In Fig. 1-7A we see the input sinewave and the pulsating DC output of a fullwave rectifier. Note that the negative halves of the sinewaves are flipped over and appear in the positive going direction. The characteristic function for the fullwave rectifier is shown in Fig. 1-7B. Because the output voltage is always positive, regardless of whether the input signal is positive or negative, the fullwave rectifier can be called an *absolute value circuit*. The output voltage will be either:

$$V_o = k|V_{in}| \tag{1-6}$$

or

$$V_o = -k|V_{in}| \tag{1-7}$$

depending upon the direction of the diodes within the circuit.

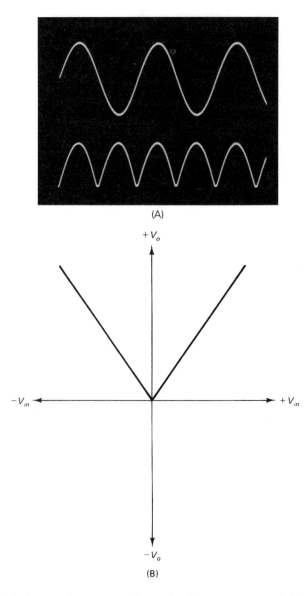

FIGURE 1-7 (A) input and output waveforms for fullwave rectifier; (B) fullwave rectifier transfer function.

While the fullwave rectifier has major applications in DC power supplies, it is the absolute value feature that makes the fullwave rectifier important for instrumentation and other related purposes.

There are several methods for creating a precise fullwave rectifier, or absolute value amplifier, some of which are shown in Fig. 1-8. The first

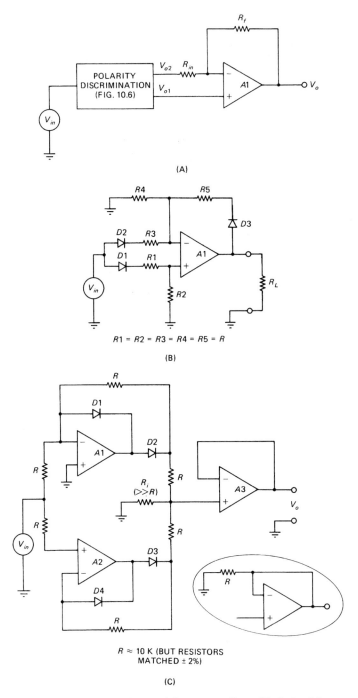

FIGURE 1-8 (A) polarity discriminator fullwave rectifier; (B) diode fullwave rectifier; (C) precise fullwave rectifier.

NONLINEAR (DIODE) APPLICATIONS OF LINEAR IC DEVICES

circuit (Fig. 1-8A) is based on the polarity discriminator circuit of Fig. 1-6. The two outputs, V_{o1} and V_{o2}, are applied to the inputs of a DC differential amplifier.

A second approach is shown in Fig. 1-8B. In this circuit a pair of oppositely connected diodes are applied to the inputs of a simple DC differential amplifier. This approach is not as well regarded because of the fact that the diodes in the input stage are not in the feedback loop, so their voltage drops (V_g) are not servoed-out.

Figure 1-8C shows the usual absolute value amplifier circuit. It consists of a pair of precise halfwave rectifier circuits connected such that their respective outputs are summed at the input of an output buffer amplifier. Amplifier $A1$ is connected as an inverting precise rectifier, while amplifier $A2$ is connected as a noninverting precise rectifier. The waveform of the signal at the summing point, i.e. the input of amplifier $A3$, is the absolute value of the input waveform.

There are two common variations of this circuit. As shown, the input circuit of $A3$ uses a resistor to ground that is very much larger than the summing resistors (R). This relationship prevents the loading of the signal by voltage divider action between R and R_i. In some modern BiFET and BiMOS amplifiers no R_i is needed because the bias current is very nearly zero.

The other alternative is to make $R_i = R$, causing the input voltage to $A3$ to be reduced to one-half by $[VR/(R+R)]$. For this case, it is necessary to make $A3$ a gain of two noninverting follower amplifier (see inset to Fig. 1-8C).

1-4.4 Zero-bound and dead-band circuits

A zero-bound circuit is one in which the output voltage is limited such that it will be non-zero for certain values of input voltage, and zero for all other input voltages. The term does not mean that the values of V_{in} are in any way constrained, but rather that there are constraints on allowable output voltages. The output of a zero-bound circuit indicates when the input signal exceeds a certain threshold, and by how much.

Figure 1-9 shows a zero-bound amplifier circuit. This circuit is based on the halfwave precise rectifier circuit of Fig. 1-3, and functions in exactly the same way except for the extra input reference current, I_{ref}. The effect of I_{ref} is to offset the trip point at which the input voltage takes effect.

To understand this circuit we can use an analysis similar to the method used before, i.e. based on the properties of the ideal operational amplifier. We know from Kirchhoff's current law (KCL) and the fact that op-amp inputs neither sink nor source current, that the following relationship is true:

$$I1 + I_{ref} + I2 = 0 \qquad (1\text{-}8)$$

or

$$I1 + I_{ref} = -I2 \qquad (1\text{-}9)$$

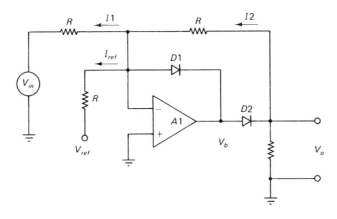

FIGURE 1-9   Zero-bound circuit.

We also know that:

$$I1 = \frac{V_{in}}{R} \quad (1\text{-}10)$$

$$I_{ref} = \frac{V_{ref}}{R} \quad (1\text{-}11)$$

$$I2 = \frac{V_o}{R} \quad (1\text{-}12)$$

Thus,

$$\frac{V_{in}}{R} + \frac{V_{ref}}{R} = -\frac{V_o}{R} \quad (1\text{-}13)$$

and after multiplying both sides by R:

$$V_{in} + V_{ref} = -V_o \quad (1\text{-}14)$$

Thus, the output voltage is still proportional to the input voltage, but is offset by the value of V_{ref}. The transfer characteristics for this circuit are shown in Fig. 1-10. In Fig. 1-10A V_{ref} is negative, while in Fig. 1-10B V_{ref} is positive. In both cases the transfer curve is offset by the reference signal potential.

Consider the operation of the circuit in Fig. 1-9 under two conditions: $V_{in} > 0$ and $V_{in} < 0$. First assume that $V_{ref} = 0$. For the positive input ($V_{in} > 0$), the output of the operational amplifier ($A1$) swings negative (the circuit is an inverter), causing diode $D2$ to be reverse biased, and $D1$ to be forward biased. Output voltage V_o is zero in this case. The output voltage is bound to zero for all values of $V_{in} > 0$.

Now consider the cases where V_{ref} is not zero. There are two cases: $V_{ref} > 0$ and $V_{ref} < 0$. Figure 1-11 shows several cases of a zero-bound circuit such as Fig. 1-9. In all of these examples a 741 operational amplifier was connected

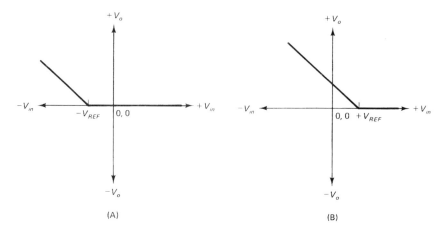

FIGURE 1-10 Transfer characteristic for different values of V_{ref}.

with a pair of 1N4148 silicon signal diodes; the value of R was selected as 22 kohms. The excitation signal was an 8 volt p–p sinewave at a frequency of 700 Hz. Figure 1-11A shows the action of the circuit without the reference voltage applied ($V_{ref} = 0$); the circuit operates as a normal precise rectifier. As will be true of all of the Fig. 1-11 examples the input sinewave is shown in the upper trace, while V_o is shown in the lower trace on the dual-beam oscilloscope.

In Fig. 1-11B $V_{ref} = -1.2$ volts. Notice the clipping action. The amplitude of the waveform is 5.2 volts base to peak. Because the input signal is 8 volts p–p, the positive peak is 4 volts peak. Thus, the baseline of the output signal voltage is at a level of $[(+4\ V) - (5.2\ V)]$, or -1.2 volts... which is

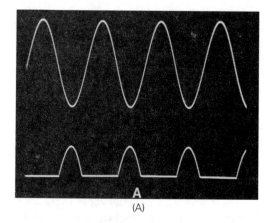

(A)

FIGURE 1-11 Oscilloscope photos showing output waveform for various conditions.

18 NONLINEAR (DIODE) APPLICATIONS OF LINEAR IC DEVICES

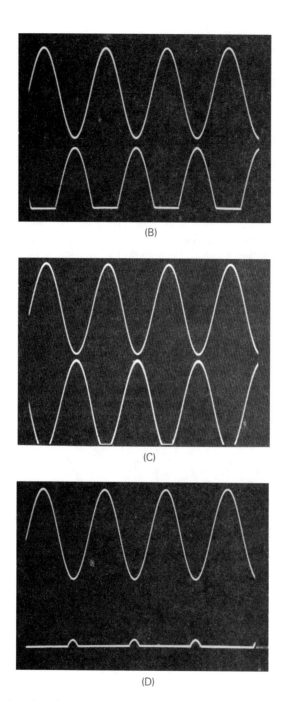

FIGURE 1-11 (continued)

the value of V_{ref}. In this circuit, the zero-bounding occurs at all negative potentials greater than -1.2 volts; only those signals more positive than this value can pass to the output of the circuit.

The waveforms in Fig. 1-11C are similar to those in Fig. 1-11B, except that the reference voltage has been increased to -3.4 volts. In this case only ($4 - 3.4$ volts), or 0.6 volts, of the negative peak is unable to pass to the output. Exactly the opposite situation is shown in Fig. 1-11D. Here the reference voltage has the same magnitude, but is reversed in polarity ($+3.4$ volts DC). Note that only the top 0.6 volts of the positive peak shows; all lower voltages are zero-bounded.

Dead-band circuits. A dead-band circuit is one in which two zero-bound circuits work together to produce a summed output. Figure 1-12A shows the transfer characteristic of such a circuit. There are two different threshold values shown in this curve. The circuit will output signals only when the input signal is less than the lower threshold ($V_{in} < -V_{th}$), or greater than the upper threshold ($V_{in} > +V_{th}$). This behavior is shown relative to a sinewave input signal. The output will be zero for all values of input signal within the shaded zone.

Keep in mind that the output voltage will not suddenly snap to a high value above the threshold potential, but rather will be equal to the difference between the peak voltage and the threshold voltage. Assuming unity gain for both reference voltage and input signal voltage, the output peaks will be $[(+V_p) - (+V_{th})]$, and $[(-V_p) - (-V_{th})]$. A waveform such as this is

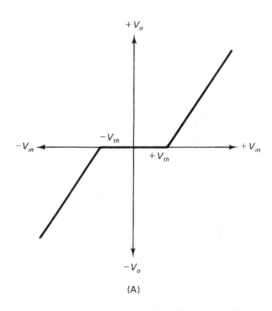

(A)

FIGURE 1-12 (A) dead-band transfer function; (B) effect of dead-band on sinewave signal; (C) input and output waveforms for dead-band circuit.

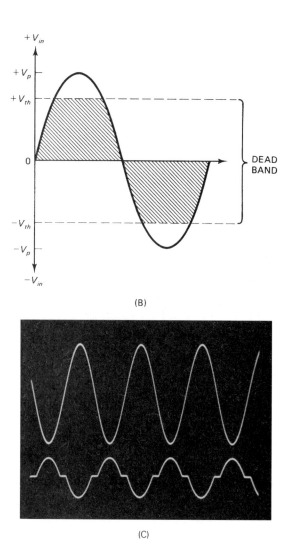

(B)

(C)

FIGURE 1-12 (continued)

shown in Fig. 1-12C. Note here that the two threshold voltages are not equal to each other, so produce different peak values.

The dead-band amplifier circuit consists of a pair of zero-bound circuits summed together (Fig. 1-13). Both zero-bound circuits are similar to Fig. 1-9. Zero-bound circuit no. 1 uses diodes in the same polarity as was used in Fig. 1-9, while zero-bound circuit no. 2 uses reverse polarity diodes (both shown in the insets to Fig. 1-13). In the case of the first circuit the $V+$ DC power supply is used as V_{ref}, while in the second the $V-$ DC supply is used as V_{ref}. In both cases the magnitude is the same, but the polarities are reversed. The difference in the threshold levels in this case is set by

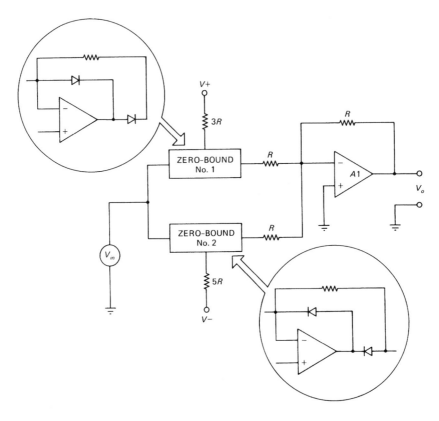

FIGURE 1-13 Zero-bound elements used to make a dead-band circuit.

using different values of reference resistor: $3R$ in the first zero-bound circuit, and $5R$ in the second zero-bound circuit. The result is the waveform of Fig. 1-12C.

1.5 PEAK FOLLOWER CIRCUITS

The peak follower is a circuit that will output the highest value input voltage that was applied to it, regardless of what the input voltage does after that point. Figure 1-14 shows the action of a typical peak follower. Input voltage V_{in} varies over a wide range. The output voltage (shown by the heavy line profile), however, always remains at the highest value reached previously, and only increases if a new peak is encountered.

A typical peak follower is shown in Fig. 1-15. This circuit is basically a noninverting halfwave precise rectifier in which a unity gain, noninverting buffer amplifier is inserted into the feedback loop between diode $D2$ and the feedback resistor. Also added to the circuit are a capacitor to hold the charge ($C1$) and a reset switch that is used to discharge the capacitor. The value of

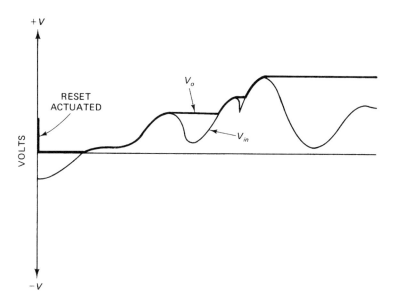

FIGURE 1-14 Action of peak follower circuit.

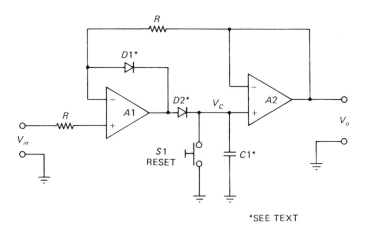

*SEE TEXT

FIGURE 1-15 Peak follower circuit.

the capacitor is selected to be small enough to allow it to charge rapidly when an input signal is applied, but large enough to not become fully charged.

When the input voltage is positive ($V_{in} > 0$), the output of $A1$ is also positive. This forces diode $D1$ to be reverse biased, and $D2$ to be forward biased. At initial turn-on, or immediately after the reset switch is closed and then re-opened, the capacitor is discharged. In this case $V_c = 0$. If a positive input voltage is applied to the input, then this potential begins to charge $C1$,

NONLINEAR (DIODE) APPLICATIONS OF LINEAR IC DEVICES   **23** causing V_c to increase to V_{in}. When the voltage across $C1$ is equal to the output voltage of the amplifier, then current flow into the capacitor ceases, and the value of V_c is at the maximum value reached by V_{in}. Because amplifier $A2$ is a noninverting, unity gain follower, the output voltage is equal to the voltage across the capacitor ($V_o = V_c$).

As long as V_{in} is equal to the capacitor voltage, or less, the capacitor voltage remains unchanged. In other words, the capacitor voltage (hence also the output voltage) remains at the previous high value. But if the input voltage should rise to a point greater than the previous peak, then $V_{in} > V_c$, so a current will flow into $C1$ and cause its voltage to reach the new level. Again, the output voltage will track the previous high value of $V_{in(max)}$. Only after reset switch $S1$ is closed momentarily will the output return to zero.

For negative input voltages diode $D2$ is reverse biased, so no current will pass either to or from the capacitor. This circuit ignores negative input potentials.

There are special precautions to take with respect to certain of the components in this circuit. Capacitor $C1$, for example, must be a very low leakage type. If there is a leakage current, which implies a shunt resistance across the capacitance, then the charge on the capacitor will bleed off with time. For the same, reason, the input impedance of amplifier $A2$ must have an extremely high input impedance. For this reason, special premium operational amplifiers are selected. BiMOS, BiFET and other very low input bias current models are preferred for this application. Also, the diode selected for $D2$ must have an extremely high reverse resistance. In other words, the leakage current that passes through $D2$ must be kept as low as possible. The reason for these precautions is to prevent the charge on $C1$ from bleeding off prematurely.

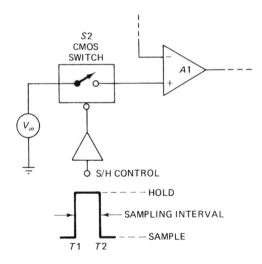

FIGURE 1-16 Sample and hold circuit.

The outward result of this circuit action is apparent 'droop' of the output voltage, V_o.

1-5.1 Sample and hold circuit

The peak follower circuit of Fig. 1-15 can be modified to form a sample and hold circuit. By adding a series switch ($S2$) at the input (Fig. 1-16) the peak follower will admit signal only at a discrete time determined by the S/H control signal. The switch is a CMOS electronic switch, and is used to allow a logic signal (as might be provided by a signal) to produce the S/H action. A similar switch can be used for $S1$ (Fig. 1-15). The correct action will first drive $S1$ closed to discharge $C1$, and then open $S1$ and close $S2$ in order to charge $C1$ with the maximum value of V_{in} reached during the $T2-T1$ sampling interval.

1.6 CLIPPER/CLAMP CIRCUITS

Clipper (or clamp) circuits are the opposite of the dead-band circuit. In these circuits the output voltage will swing at will around zero provided that the input signal does not exceed a certain predetermined threshold. Either the positive peak, the negative peak, or both, will be limited to a certain clamped value. Figure 1-17 shows a typical example. These waveforms were taken in an inverting follower circuit with a gain of (-100 kohms/22 kohms), or about 4.6. In Fig. 1-17A the input signal (upper trace) is close to, but below, the critical value. The output voltage is, therefore, unclipped on the positive peak and only moderately clipped on the negative peak. In Fig. 1-17B, however, the input signal is considerably increased, but the amplifier has no further

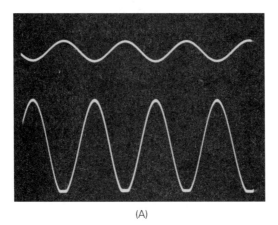

(A)

FIGURE 1-17 (A) input and output waveforms showing little or no clipping; (B) input and output waveforms showing distinct clipping; (C) clipping amplifier transfer function.

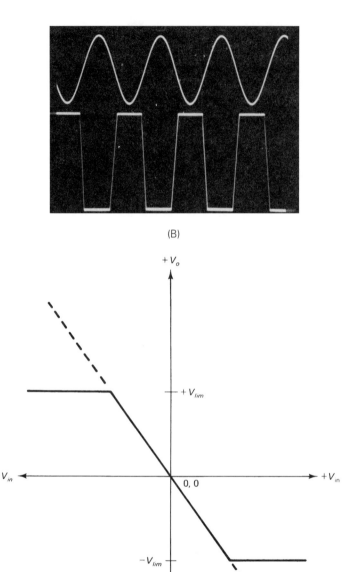

FIGURE 1-17 (continued)

output voltage to offer. At a result, the peaks of the amplifier output are clamped to a value determined by the DC power supply potential applied to the amplifier.

Figure 1-17C shows a transfer characteristic for an inverting clipper. Output voltage V_o is allowed to swing only between the lower and upper limits ($-V_{LIM}$ and $+V_{LIM}$, respectively). The dotted lines represent the output voltage that would exist in the absence of limiting.

It is not generally satisfactory to limit the DC power supply potentials in order to achieve clipping. The usual procedure is to use the full power supply potential, but to limit amplifier output voltage by certain circuit methods. Figure 1-18 shows one popular, but largely unsatisfactory, method for limiting the output swing. The feedback resistor is shunted by a pair

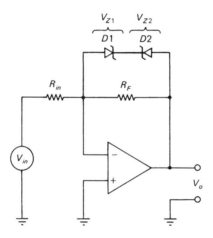

FIGURE 1-18 Zener diode op-amp clipper circuit.

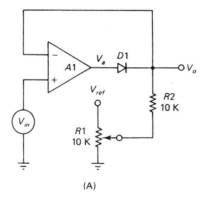

(A)

FIGURE 1-19 (A) Variable clipping point clipper circuit; (B) input and output waveforms shows that the results are not very satisfactory.

NONLINEAR (DIODE) APPLICATIONS OF LINEAR IC DEVICES **27**

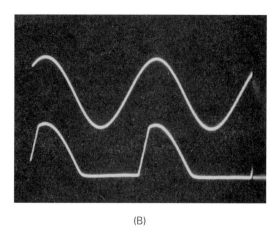

(B)

FIGURE 1-19 (continued)

of back to back zener diodes. On positive output voltages, D2 is forward biased and D1 is reverse biased. As long as $+V_o$ is less than V_{Z1} plus the forward drop across D2 (about 0.6 volts), it follows the dictates of the usual transfer equation ($V_o = V_{in}(R_f)/R_{in}$). At values greater than $V_{Z1} + 0.6$ volt, however, the output is clamped. Similarly with negative output potentials. On negative swings of the output voltage the clamp occurs at $-V_{Z2} - 0.6$ volts.

The circuit of Fig. 1-18 is not well regarded because it requires a relatively heavy signal level to sustain the zener diodes in the avalanche condition. A somewhat different approach is shown in Fig. 1-19A. In this case the amplifier drives a diode (D1). In the absence of V_{ref}, the diode will be forward biased for positive values of V_{in}. But when V_{ref} is applied, it will reverse bias D1 and prevent an output voltage until V_a overcomes the reference potential. The results of this circuit are also not very satisfactory (see Fig. 1-19B).

A more satisfactory approach is shown in Fig. 1-20. A diode bridge (D1–D4) is inserted into the feedback loop of amplifier A1. As a result, some of the nonlinearities that afflict the other circuits are servoed-out of this circuit by the $1/\beta A_v$ factor (in which β is the transfer equation of the feedback network). There are two limiting conditions for this circuit:

1. To set $+V_{LIM}$:

$$+V_{LIM} = \frac{(+V_{ref})R4R5}{R4R5 + R3R5 + R3R4} \qquad (1\text{-}15)$$

2. To set $-V_{LIM}$:

$$-V_{LIM} = \frac{(-V_{ref})R4R5}{R4R5 + R3R5 + R3R4} \qquad (1\text{-}16)$$

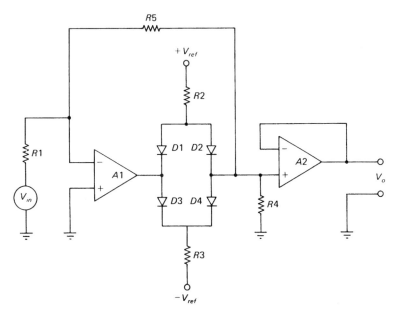

FIGURE 1-20 Best approach to making a clipper circuit.

1.7 SUMMARY

1. The PN junction diode is imperfect. It behaves in a nonlinear logarithmic manner between zero and a certain critical voltage, V_g, that is 0.2 to 0.3 for Ge diodes, and 0.6 to 0.7 volts for Si diodes.
2. A precise rectifier is an operational amplifier circuit that produces a nearly ideal transfer characteristic by servoing-out the errors in the feedback loop.
3. A zero-bound circuit is one in which a reference voltage limits a portion of the transfer characteristic to zero. A dead-band circuit combines two zero-bound circuits to produce a characteristic that limits output voltage to zero for a zone around zero.
4. A peak follower circuit is one that stores the input voltage in a capacitor, and always outputs the highest value of that input voltage even if V_{in} subsequently drops.
5. A sample and hold circuit is a peak follower type of circuit that is switched on and off by a control signal. The output voltage is the highest level attained by the input signal during the sampling interval.
6. A clipper circuit is one that limits the maximum attainable value of the output voltage. It is the opposite of a dead-band circuit.

1.8 RECAPITULATION

Now return to the objectives and Pre-quiz questions at the beginning of the chapter and see how well you can answer them. If you cannot answer certain questions, place a check mark to each and review the appropriate parts of the text. Next, try to answer the questions and work the problems below, using the same procedure.

1.9 STUDENT EXERCISES

1. Design and build an inverting halfwave precise rectifier using an operational amplifier. Examine the input and output signals on an oscilloscope. Alternatively, plot the transfer characteristic for data sampled with a voltmeter.
2. Design and build an inverting zero-bound circuit. Select different values of V_{ref}, both positive and negative.
3. Design and build a dead-band circuit: (a) with equal thresholds for positive and negative peaks, (b) with asymmetrical peaks.
4. Design and build a clipper based on Fig. 1-20.
5. Design and build a clipper based on Fig. 1-19.

1.10 QUESTIONS AND PROBLEMS

1. Draw the transfer characteristic of ideal and practical PN junction diodes.
2. Draw the circuit for the inverting halfwave precise diode circuit. Label all components and describe circuit action in your own words.
3. Sketch the output waveforms for a sinewave input for a precise diode circuit such as Fig. 1-3.
4. A 741 operational amplifier has an open-loop gain of 250 000 and a gain–bandwidth product of 1.25 MHz. Calculate the smallest input signal that a precise diode based on this amplifier will accept. Assume that silicon diodes are used and that $V_g = 0.65$ volts.
5. A precise diode is used as the demodulator in an amplitude modulated carrier amplifier. Calculate the maximum frequency of the carrier frequency that will still result in low-error demodulation. Assume a closed-loop voltage gain of 10 and a gain–bandwidth product of 1.25 MHz.
6. What would happen in the problem above if the gain was increased to 1000?
7. What is the minimum slew rate required for a precise diode used as an AM detector if the carrier frequency is 1000 Hz, the gain is 10, and the input voltage is 4 volts p–p?
8. In Fig. 1-20 all resistors are equal. $+V_{ref} = 12$ volts, and $-V_{ref} = -12$ volts. Calculate the values of the limit potentials.
9. Why are ideal diode rectifier circuits preferred in some applications than actual PN junction diode rectifiers?
10. What are the nominal junction voltages of: (a) germanium PN junction diodes and (b) silicon PN junction diodes?
11. How does the 'ideal' or 'precise' rectifier circuit remove the normal junction voltage drop inherent in PN junction diodes?

12. The minimum signal allowed in a silicon diode based precise rectifier when the open-loop voltage gain is 300 000 is _____ .
13. Calculate the minimum signal that can be accurately rectified in a precise rectifier circuit in which the open-loop gain is 2000. Assume that a silicon diode is used.
14. What is the maximum allowable carrier frequency for a precise rectifier AM demodulator if the gain–bandwidth product of the op-amp is 4.5 MHz and the voltage gain is 15?
15. Calculate the values of $C1$ in Fig. 1-3 if a 2 kHz carrier is used and both resistors are 12 kohms. Assume a peak signal voltage of 1.25 volts.
16. Sketch the circuit of a polarity discriminator circuit.
17. Sketch the circuit of a fullwave precise rectifier.
18. Sketch the circuit of a zero-bound circuit.
19. Sketch the circuit of a dead-band circuit.
20. Sketch the circuit of a peak follower circuit.
21. Sketch the circuit of a sample and hold circuit.
22. Calculate the boundary limits for a circuit such as Fig. 1-20 if $-V_{ref} = -6$ Vdc, $+V_{ref} = +9$ Vdc and all resistors are 10 kohms.

CHAPTER 2

Signal processing circuits

OBJECTIVES

1. Learn how active differentiator and integrator circuits work.
2. Learn how active logarithmic and anti-log amplifiers work.
3. Understand the practical applications of analog signals processing circuits.

2-1 PRE-QUIZ

These questions test your prior knowledge of the material in this chapter. Try answering them before you read the chapter. Look for the answers (especially those you answered incorrectly) as you read the text. After you have finished studying the chapter try answering these questions again, and those at the end of the chapter (see Section 2-11).

1. An *integrator* circuit finds the area under a curve over an interval, so for a time-varying analog signal it finds the _____ _____ of the signal.
2. Draw the simplified circuit for a logarithmic amplifier.
3. An operational amplifier Miller integrator circuit is connected to a DC input signal of 10 mV DC. If the resistor is 1 megohm and the capacitor is 1 µF, how long does it take the output voltage to reach 1 Vdc?
4. What is the gain of a Miller integrator that uses a feedback capacitance of 100 pF and an input resistance of 100 kohms?

2-2 INTRODUCTION

Electronic instrumentation usually requires at least some amplification or processing of analog electrical signals. Even in an era of massive computerization of such instruments, there is still need for the analog subsystem in the front-end. It might be necessary, for example, to boost an analog signal to the point where it can be input to the A/D converter connected to a computer. In addition to this simple scaling function it is also often desirable to do some of the signals processing in the analog subsystem. While this statement may seem unrealistic to computer oriented people, it is nonetheless often a reasonable trade-off in many situations. There may be a situation in which computer hardware or timeline constraints makes it less costly to use a simple analog circuit. It is often asserted that the computer solution is 'better' than the analog circuit solution. While this claim may be true much of the time, it is not universally true. As device manufacturer catalogs attest, analog signals processing is far from dead — it is alive, healthy, larger than ever, and still growing.

In this chapter we are going to take a look at standard laboratory amplifiers that are used for certain signals acquisition chores in electronic instrumentation. In addition, we will examine certain linear IC circuits used for analog signals processing.

The term 'laboratory amplifiers' describes a wide range of instruments of many and varying capabilities. Although some are quite complex, many of them can be designed using simple linear IC devices. Some of these instruments are categorized according to several schemes. For example, we can divide them according to input coupling method: DC versus AC. In the case of AC amplifiers there is often a frequency response characteristic that will take some of the burden of filtering in the system. We can also categorize the amplifiers according to gain:

Low gain	1 to 100
Medium gain	101 to 1000
High gain	>1000

These ranges are commonly accepted, but because they are popularly established rather than established through formal industry standards they may vary somewhat from one manufacturer to another. The three categories nonetheless serve as a reasonable context for our discussion.

Some amplifiers carry names that represent certain special applications. For example, the *biopotentials amplifier* is used to acquire natural electrical signals from living things and, because of certain practical problems, tend to have very high input impedances and certain other characteristics that establish the class.

Laboratory amplifiers can be either free-standing models, or part of a plug-in mainframe data logging or instrumentation systems. In the sections below we will discuss some of the special forms of laboratory amplifier that may be useful in certain specific cases.

2-3 CHOPPER AMPLIFIERS

One of the unfortunate characteristics of simple DC amplifiers is that they may be noisy and possess a certain inherent *thermal drift* of both gain and DC offset baseline (especially the latter). In low and medium gain applications these problems are less important than in high gain amplifiers, especially in the lower regions of those gain ranges. As gain increases, however, these problems loom much larger. For example, a drift of 50 µV/°C in a ×100 medium gain amplifier produces an output voltage change of

$$(50 \ \mu V/°C) \times 100 = 5 \ mV/°C$$

A drift of 5 mV/µV/°C is certainly tolerable in most low gain circuits, but in a X20 000 high gain amplifier the output voltage would escalate to:

$$(50 \ \mu V/°C) \times 20\,000 = 1 \ V/°C$$

This level of drift will obscure any real signals in a very short period of time.

Similarly, noise can be a problem in high-gain applications, where it had been negligible in most low to medium gain applications. Operational amplifier noise is usually specified in terms of nanovolts of noise per square root hertz (i.e. NOISE(rms) = $nV/(\sqrt{Hz})$). A typical low-cost operational amplifier has a noise specification of 100 $nV/(\sqrt{Hz})$, so at a bandwidth of 10 kHz the noise amplitude will be

$$\text{NOISE}_{RMS} = 100 \ nV \times \sqrt{10\,000}$$

$$\text{NOISE}_{RMS} = 100 \ nV \times 100$$

$$\text{NOISE}_{RMS} = 10\,000 \ nV = 0.00001 \ \text{volts}$$

In a ×100 amplifier, without low-pass filtering, the output amplitude will be only 1 mV, but in a ×100 000 amplifier it will be 1 volt.

A circuit called a *chopper amplifier* can solve both problems because it makes use of a relatively narrowband AC-coupled amplifier in which the advantages of feedback can be optimized. The drift problem is reduced significantly by two properties of AC amplifiers. One property is the inability to pass low frequency (i.e. near-DC) changes such as those caused by drift. To the amplifier, drift looks like a valid low-frequency (sub-hertzian) signal. The other property is the ability to regulate the amplifier through the use of negative feedback.

Many low-level analog signals are very low frequency, i.e. in the DC to 30 Hz range (for example, human electrocardiogram signals have frequency components down to 0.05 Hz), so will not pass through a narrowband AC amplifier. The solution to this problem is to chop the signal so that it passes through the AC amplifier, and then to demodulate the amplifier output signal to recover the original waveshape, but at a higher amplitude.

Figure 2-1 shows a block diagram of the basic chopper amplifier circuit. The traditional chopper mechanism is a vibrator-driven SPDT switch (S1) connected so that it alternately grounds first the input and then the output of the AC amplifier. An example of a chopped waveform is shown in Fig. 2-2. A low-pass filter following the amplifier will filter out any residual chopper 'hash' and any miscellaneous noise signals that may be present.

Most old-fashioned mechanical choppers used a chop rate of either 50/60 Hz or 400 Hz, although 100 Hz, 200 Hz and 500 Hz choppers are also found. The main criterion for the chop rate is that it be at least twice the highest component frequency that is present in the input waveform (Nyquist's sampling criterion).

A differential chopper amplifier is shown in Fig. 2-3. In this circuit an input transformer with a center-tapped primary is used. One input terminal is connected to the transformer primary center-tap, while the other input terminal is switched back and forth between the respective ends of the transformer primary winding. A synchronous demodulator following the AC amplifier detects the signal, and restores the original, but now amplified, waveshape. Again, a low-pass filter smooths out the signal.

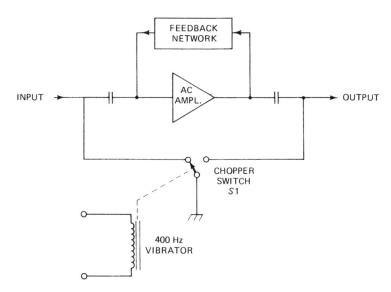

FIGURE 2-1 Chopper amplifier using electromechanical vibrator as the input/output switcher.

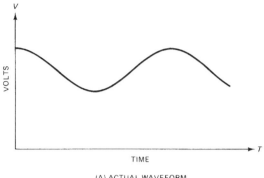
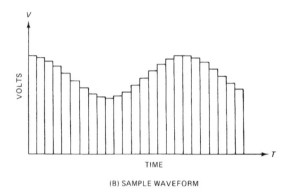

FIGURE 2-2 (A) actual waveform; (B) sampled waveform.

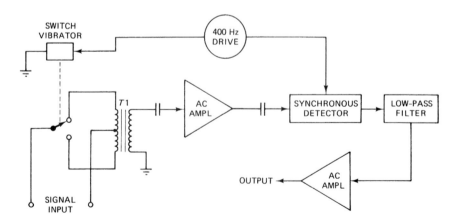

FIGURE 2-3 Differential input chopper amplifier block diagram.

36 SIGNAL PROCESSING CIRCUITS

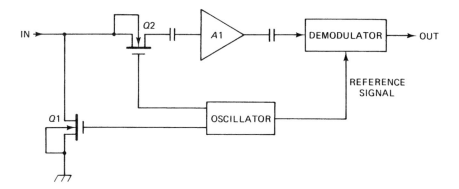

FIGURE 2-4 MOSFET transistors used as input switchers.

The modern chopper amplifier may not use mechanical vibrator switches as the chopper. A pair of CMOS or JFET electronic switches driven out of phase with each other will perform exactly the same function. Other electronic switches used in commercial chopper amplifiers include PIN diodes, varactors, and optoisolators. Figure 2-4 shows a modern electronically chopped amplifier that can be obtained in either IC or hybrid form.

Chopper amplifiers limit the noise because of the both the low-pass filtering required and because of the fact that the AC amplifier frequency response can be set to a narrow passband around the chopper frequency.

At one time the chopper amplifier was the only practical way to obtain low drift in high gain situations. Modern IC and hybrid amplifiers, however, have such improved drift properties that no chopper is needed (especially in the lower end of the 'high gain' range). The Burr-Brown OPA-103 is a monolithic IC operational amplifier in a TO-99 eight-pin metal can package. This low-cost device exhibits a drift characteristic of 2 µV/°C. In addition, there are amplifiers available (especially in hybrid form) that are actually electronic chopper stabilized, but to the outside world the device simply looks like a low-drift amplifier. The Burr-Brown 3271/25 device is an example that exhibits a drift characteristic of 0.1 µV/°C.

2-4 CARRIER AMPLIFIERS

A *carrier amplifier* is any type of signal processing amplifier in which the signal carrying the desired information is used to modulate another (usually higher frequency) signal, i.e. a 'carrier' signal. The chopper amplifier is considered by many to fit this definition, but is usually regarded as a unique type in its own right. The three principal carrier amplifiers are the *DC-excited*, *AC-excited* and *pulse-excited* varieties.

Figure 2-5 shows a DC-excited carrier amplifier. A Wheatstone bridge transducer provides the input signal, and is excited by a DC potential, V.

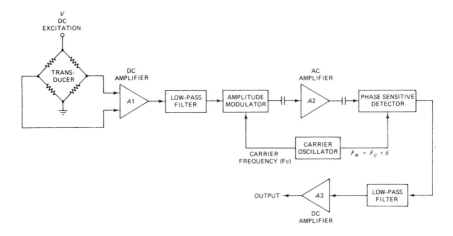

FIGURE 2-5 DC-excited carrier amplifier.

The output of the transducer is a low-level DC voltage that varies with the value of the stimulating parameter. The transducer signal is usually of very low amplitude, and may be noisy. An amplifier increases the signal amplitude, and a low-pass filter removes much of the noise. In some models the first stage is actually a composite of these two functions, being essentially a filter with gain.

The signal at the output of the amplifier-filter section is used to *amplitude modulate* a carrier signal. Typical carrier frequencies range from 400 Hz to 25 kHz, with 1 kHz and 2.4 kHz being very common. The signal frequency response of a carrier amplifier is a function of the carrier frequency, and is usually (at maximum) one-fourth of that carrier frequency. A carrier frequency of 400 Hz, then, is capable of signal frequency response of 100 Hz, while the 25 kHz carrier will support a frequency response of 6.25 kHz. Further amplification of the signal is provided by an AC amplifier.

The key to the performance of any carrier amplifier is the *phase sensitive detector* (PSD) that demodulates the amplified AC signal. Envelope detectors, while very simple and low-cost, suffer from an inability to discriminate between the real signals and certain spurious signals.

The advantages of the PSD include the fact that it rejects signals not of the carrier frequency, and certain signals that are of the carrier frequency. The PSD, for example, will reject even harmonics of the carrier frequency, and those components that are out of phase with the reference signal. The PSD will, however, respond to odd harmonics of the carrier frequency. Some carrier amplifiers seem to neglect this problem altogether. But in some cases, manufacturers will design the AC amplifier section to be a bandpass amplifier with a response limited to $F_c \pm F_c/4$. This response will eliminate any third or higher order odd harmonics of the carrier frequency before they reach the PSD. It is then only necessary to assure that the reference signal has acceptable purity (harmonic distortion and phase noise).

An alternative, but very common, form of carrier amplifier is the AC-excited carrier amplifier circuit shown in Fig. 2-6A. In this circuit the transducer is AC-excited by the carrier signal, eliminating the need for the amplitude modulator. The small AC signal from the transducer is amplified and filtered before being applied to the PSD circuit. Again, some designs use

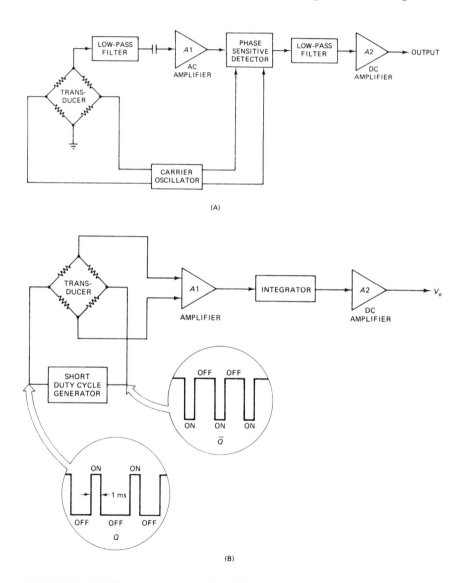

FIGURE 2-6 (A) AC-excited carrier amplifier; (B) pulse-excited carrier amplifier.

SIGNAL PROCESSING CIRCUITS

a bandpass AC amplifier to eliminate odd harmonic response. This circuit allows adjustment of transducer offset errors in the PSD circuit instead of in the transducer, by varying the phase of the reference signal.

Figure 2-6B shows the block diagram to a transducer amplifier that uses pulsed excitation for the Wheatstone bridge. A short duty cycle pulse generator produces either monopolar or (in the case shown) bipolar pulses that are used to provide the excitation potential. The advantage of the pulsed method is that the short duty cycle limits the amount of power dissipated in the transducer resistance elements, and therefore reduces the effects of transducer self-heating on the thermal drift. Several methods are used to demodulate the pulse waveform at the output of amplifier $A1$. A Miller integrator, which finds the time average of the signal, will create a DC potential that is proportional to the amplified transducer output signal. Alternative schemes use CMOS or other electronic switches to demodulate the signal.

Another advantage of the pulsed scheme is that an *amplifier drift cancellation* circuit can easily be implemented. Switching is provided that shorts together the input of $A1$ during the 'off time' of the pulse, and connects a capacitor to the output of $A1$. When the capacitor is charged, it can be connected as an offset null potential to amplifier $A2$ during the 'on time' of the pulse. Tektronix, Inc. once used a similar scheme as a *baseline stabilization* method in a medical patient monitor oscilloscope.

2-5 LOCK-IN AMPLIFIERS

The amplifiers discussed so far in this chapter produce relatively large amounts of noise, and will respond to any noise present in the input signal. They suffer from the usual shot noise, thermal noise, H-field noise, E-field noise, ground loop noise, and so forth that affects all amplifiers. The noise at the output is directly proportional to the square root of the circuit bandwidth. The *lock-in amplifier* is a special case of the carrier amplifier in which the bandwidth is very narrow. Some lock-in amplifiers use the carrier amplifier circuit of Fig. 2-6, but with an input amplifier with a very high Q bandpass characteristic. The carrier frequency will be between 1 kHz and 200 kHz. The lock-in principle works because the information signal is made to contain the carrier frequency in a way that is easy to demodulate and interpret. The AC amplifier accepts only a narrow band of frequencies centered about the carrier frequency. The narrowness of the amplifier bandwidth, which makes possible the improved signal-to-noise ratio, also limits the lock-in amplifier to very low frequency input signals. Even then, it is sometimes necessary to integrate (i.e. time-average) the signal in order to obtain the needed data.

Lock-in amplifiers are capable of reducing the noise, and retrieving signals that are otherwise buried below the noise level. Improvements of up to 85 dB are relatively easily obtained, and up to 100 dB reduction is possible if cost is less of a factor.

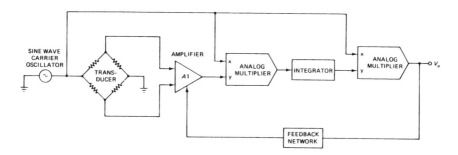

FIGURE 2-7 Autocorrelation amplifier.

There are actually several different forms of lock-in amplifier available. The type discussed above is the simplest type. It is merely a very narrowband version of the AC-excited carrier amplifier. The lock-in amplifier of Fig. 2-7, however, uses a slightly different technique. It is called an *autocorrelation amplifier*. The carrier is modulated by the input signal, and then integrated. The output of the integrator is demodulated in a product detector circuit. The circuit of Fig. 2-7 produces very low output voltages for input signals that are not in-phase with the reference signal, but produces relatively high output voltages at the proper input frequency and phase.

2-6 ELECTRONIC INTEGRATORS AND DIFFERENTIATORS

There are two processes in mathematics that are very important to electronic instrumentation and signals processing: *integration* and *differentiation*. These processes are inverses of each other. In other words, a function that is first integrated and then differentiated, returns to the original function. A similar relationship occurs when a function is differentiated and is then integrated. Such is the normal nature of mathematically inverse processes.

These processes are also seen elsewhere in electronics, but sometimes under different names. The differentiator is sometimes called a *rate of change circuit*, or, if the time constant is correct, a *high-pass filter*. Similarly, the integrator might be called a *time averager circuit* or *low-pass filter*.

Although a rigorous definition of the processes of integration and differentiation is beyond the scope of this book, a simple introduction will serve to illustrate how these circuits work. Consider first the process of integration.

Figure 2-8 shows an analog voltage waveform that varies as a function of time ($V = F(t)$). How do you find the area under the curve between $T1$ to $T2$? If the voltage is constant over the range $T1$ to $T2$ (as in the case of DC), then you can simply take the product $V \times (T2 - T1)$. But V is not always constant over the interval of interest. However, if you break the curve into tiny intervals ($T_b - T_a$ in the inset to Fig. 2-8) then the voltage change over that short interval is small enough that we may consider it essentially

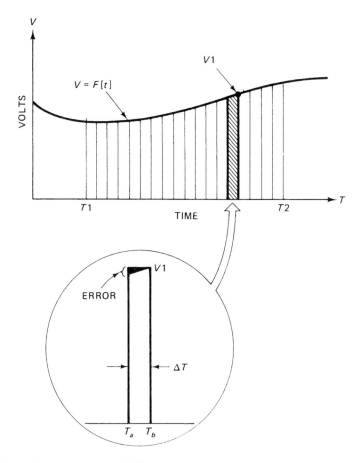

FIGURE 2-8 Rectangles method of integration.

constant. We can then approximate the area under the curve over that short interval as $V1 \times (T_b - T_a)$, or, simply, $V1\Delta T$. Assuming that the time width of all intervals is the same, the total area under the curve $V = F(t)$ over the interval $T1$ to $T2$ is approximated by summing the individual interval areas:

$$AREA = \sum_{i=1}^{n} V_i \Delta T \qquad (2\text{-}1)$$

where:
   V_i is the voltage at the ith interval
   ΔT is the time interval common to all samples

The validity of the approximation increases as the time interval decreases. When ΔT becomes infinitesimally small, the approximation becomes exact, and the notation used for Eq. (2-1) changes to

$$AREA = \frac{1}{T} \int_{t1}^{t2} V \, dt \qquad (2\text{-}2)$$

where:
  V is the voltage as a function of T
  ΔT is $t2 - t1$

Multiplying by the factor $1/T$ reduces the result to the *time average* of the function.

In Fig. 2-9 a voltage represents a pressure transducer output, in this particular case the output of a human blood pressure transducer. Such transducers are common in hospital intensive care units. Notice that the pressure voltage varies with time from a low ('diastolic') to a high ('systolic') between $T1$

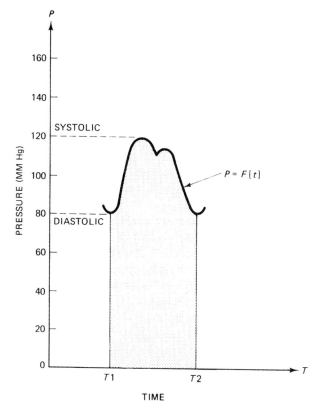

FIGURE 2-9 Human arterial blood pressure waveform. Mean arterial pressure is represented by area under the curve. The heart beats at T1 and T2.

and T2 (which interval represents one complete cardiac cycle). If you want to know the *mean arterial blood pressure* (MAP), then you must find the *area under the pressure versus time curve* over one cardiac cycle.

An electronic integrator circuit serves to compute the time-average of the analog voltage waveform that represents the time-varying arterial blood pressure. In an electronic blood pressure monitoring instrument, a voltage serves to represent the pressure. If, for example, a scaling factor of 10 mV/mmHg is used (as is commonly the case in medical devices), then a pressure of 100 mmHg is represented by a potential of 1000 mV, or 1.000 volt. This voltage will vary over the range 800 mV to 1200 mV for the case shown in Fig. 2-9 (pressure varies from 80 mmHg to 120 mmHg).

Differentiation is the mathematics of finding the *derivative* of a curve, which is merely its time rate of change. For the simplest case, a straight line $(Y = mX + b)$ as shown in Fig. 2-10A, the *derivative* is simply the *slope* of the line, or $(Y2 - Y1)/(X2 - X1)$. In this case, we usually write the expression for the slope with the Greek letter 'delta' (i.e $\Delta Y/\Delta X$) to indicate a small change in X and a small change in Y. For the case of a straight line, the derivative is simple to calculate. But in electronics one very frequently encounters situations where the line is not so straight, as in Fig. 2-10B where a voltage varies with time. If we want to know the *instantaneous rate of change* (i.e. the derivative) at a specific point, then we can find the *slope of a line tangent to that point*.

Electronic integrators and differentiators affect signals in different ways. Figure 2-11 shows the example of a squarewave (Fig. 2-11A) applied to the inputs of an integrator and differentiator. The integrator output is shown at Fig. 2-11B while the differentiator output is shown at Fig. 2-11C.

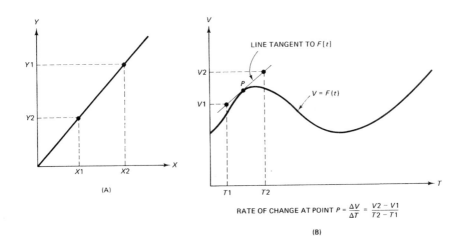

FIGURE 2-10 (A) Derivative of a straight line is the slope of the line; (B) derivative of a point on a curved line is the slope of a line tangent to that point.

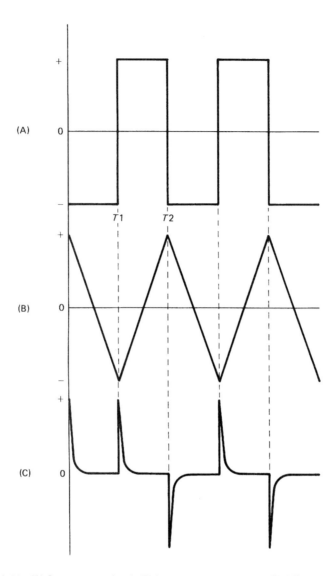

FIGURE 2-11 (A) Squarewave signal; (B) integrated squarewave; (C) differentiated squarewave.

First consider the operation of the integrator circuit. The integrator output waveform in Fig. 2-11B shows a constant positive-going slope between $t1$ and $t2$. The steepness of the slope is dependent upon the amplitude of the input squarewave, but the line is linear. You can see from curve B in Fig. 2-11 that the squarewave into the integrator produces a triangle waveform.

Now consider the operation of the differentiator circuit (see output waveform Fig. 2-11C). At time $T1$ the squarewave makes a positive-going transition to maximum amplitude. At this instant it has a very high rate of change, so the output of the differentiator is very high (see waveform in Fig. 2-11C at $T1$). But then the amplitude of the input signal reaches maximum and remains constant until $T2$, when it drops back to its previous value. Thus, the differentiator will produce a sharp positive-going spike at $T1$ and a sharp negative-going spike at $T2$. In an ideal circuit there will no transition between these states, but in real circuits there is an exponential transition that is proportional to the RC time constant of the circuit and the risetime of the waveform. Differentiator output spikes are frequently used in circuits such as timers and zero-crossing detectors.

If a sinewave is applied to the inputs of either integrators or differentiators, then the result is a sinewave output that is shifted in-phase 90°. The principal difference between the two forms of circuit is in the direction of the phase shift. Such circuits are frequently used to provide quadrature or 'sine–cosine' outputs from a sinewave oscillator.

2-6.1 RC integrator circuits

The simplest form of integrator and differentiator are simple resistor and capacitor (RC) circuits, such as shown in Fig. 2-12. The integrator is shown in Fig. 2-12A, while the differentiator is in Fig. 2-12B. The integrator consists of a resistor element in series with the signal line, and a capacitor across the signal line. The differentiator is just the opposite, the capacitor is in series with the signal line while the resistor is in parallel with the line. These circuits are also known as *low-pass* and *high-pass RC filters*, respectively. The

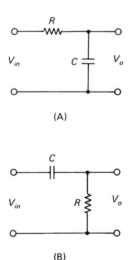

FIGURE 2-12 (A) *RC* integrator circuit; (B) *RC* differentiator circuit.

low-pass case (integrator) has a −6 dB/octave falling characteristic frequency response, while the high-pass case (differentiator) has a +6 dB/octave rising frequency response.

The operation of the integrator and differentiator is dependent upon the time constant of the RC network (i.e. $R \times C$). The integrator time constant is set long (i.e. >10X) compared with the period of the signal being integrated, while in the differentiator the RC time constant is short (i.e. <1/10X) compared with the period of the signal. Several integrators can be connected in cascade in order to increase the time-averaging effect, or increase the slope of the frequency response fall off.

2-6.2 Active differentiator and integrator circuits

The operational amplifier makes it relatively easy to build high quality active integrator and differentiator circuits. Previously, one had to construct a stable, drift-free, high-gain transistor amplifier for this purpose. Figure 2-13 shows the basic circuit of the operational amplifier differentiator. Again the RC elements are used, but in a slightly different manner. The capacitor is in series with the op-amp inverting input, while the resistor is the op-amp feedback resistor.

Analysis of the circuit to derive the transfer function follows a procedure similar to that followed for inverting and noninverting followers earlier in this book.

From Kirchhoff's current law (KCL):

$$I2 = -I1 \qquad (2\text{-}3)$$

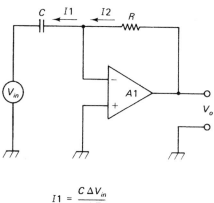

$$I1 = \frac{C \, \Delta V_{in}}{\Delta T}$$

$$I2 = \frac{V_o}{R}$$

FIGURE 2-13 Op-amp active differentiator circuit.

From basic passive circuit theory, including Ohm's law:

$$I1 = \frac{C \Delta V_{in}}{\Delta t} \qquad (2\text{-}4)$$

$$I2 = \frac{V_o}{R} \qquad (2\text{-}5)$$

Substituting Eqs (2-4) and (2-5) into Eq. (2-3):

$$\frac{V_o}{R} = -\frac{C \Delta V_{in}}{\Delta t} \qquad (2\text{-}6)$$

or, with the terms rearranged:

$$V_o = -RC \frac{\Delta V_{in}}{\Delta t} \qquad (2\text{-}7)$$

where:
 V_o and V_{in} are in the same units (volts, millivolts, etc.)
 R is in ohms (Ω)
 C is in farads (F)
 t is in seconds (s)

Equation (2-7) is a mathematical way of saying that output voltage V_o is equal to the product of the RC time constant, the derivative of input voltage V_{in} with respect to time (the '$\Delta V_{in}/\Delta t$' part). Since the circuit is essentially a special case of the familiar inverting follower circuit, the output is inverted, hence the negative sign.

Figure 2-14 shows the classical operational amplifier version of the *Miller integrator circuit*. Again, an operational amplifier is the active element, while a resistor is in series with the inverting input and a capacitor is in the feedback loop. Notice that the placement of the capacitor and resistor elements are exactly opposite in both the RC and operational amplifier versions of integrator and differentiator circuits. In other words, the RC elements reverse

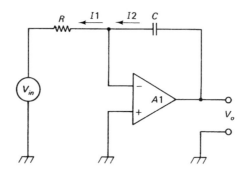

FIGURE 2-14 Op-amp active integrator circuit.

roles between Figs 2-13 and 2-14. That fact will tell the astute student a quite a bit regarding the nature of integration and differentiation.

The output of the integrator is dependent upon the input signal amplitude and the RC time constant. The transfer function for the Miller integrator is derived in a manner similar to that of the differentiator:

$$I2 = -I1 \tag{2-8}$$

From Ohm's law:

$$I1 = \frac{V_{in}}{R} \tag{2-9}$$

and, from our knowledge of how capacitors work:

$$I2 = C\frac{\Delta V_o}{\Delta t} \tag{2-10}$$

Substituting Eqs (2-9) and (2-10) into Eq. (2-8):

$$\frac{C\Delta V_o}{\Delta t} = \frac{-V_{in}}{R} \tag{2-11}$$

Integrating both sides:

$$\int \frac{C\Delta V_o}{\Delta t}\,dt = \int \frac{-V_{in}}{R}\,dt \tag{2-12}$$

$$CV_o = \int \frac{-V_{in}}{R}\,dt \tag{2-13}$$

Collecting and rearranging terms:

$$CV_o = \frac{-1}{R}\int V_{in}\,dt \tag{2-14}$$

$$V_o = \frac{-1}{RC}\int V_{in}\,dt \tag{2-15}$$

And accounting for initial conditions:

$$V_o = \frac{-1}{RC}\int V_{in}\,dt + K \tag{2-16}$$

where:
  V_o and V_{in} are in the same units (volts, millivolts)
  R is in ohms (Ω)
  C is in farads (F)
  t is in seconds (s)

This expression is a way of saying that the output voltage is equal to the time-average of the input signal plus some constant K, which is the voltage that may have been stored in the capacitor from some previous operation (often zero in electronic applications).

2-6.3 Practical circuits

The circuits shown in Figs 2-13 and 2-14 are classic textbook circuits. Unfortunately, they don't work very well in some practical cases. The problem is that these circuits are too simplistic because they depend upon the properties of ideal operational amplifiers. Unfortunately, real op-amps fall far short of the ideal in several important ways that affect these circuits. In real circuits, differentiators may 'ring' or oscillate, and integrators may saturate from their tendency to integrate bias currents and other inherent DC offsets very shortly after turn-on.

There is another problem with this kind of circuit, and it magnifies the problem of saturation. Namely, the integrator circuit of Fig. 2-14 has a very high gain with certain values of R and C. The *voltage gain* of this circuit is given by the term $-1/RC$ which, depending on the values selected for R and C, can be quite high.

EXAMPLE 2-1

Calculate the gain of a Miller integrator circuit that uses a 0.01 µF capacitor and a 10 000 ohm resistor.

Solution

$$A_v = -1/RC$$
$$A_v = -1/(10\,000 \text{ ohms})(0.000\,000\,01 \text{ farads})$$
$$A_v = -1/0.0001 = -10\,000 \qquad \blacksquare$$

In other words, with a gain of $-10\,000$, a $+1$ volt applied to the input will want to produce a $-10\,000$ volt output. Unfortunately, the operational amplifier output is limited to the range of approximately -10 volts to -20 volts, depending upon the device and the applied $V-$ DC power supply voltage. For this case, the operational amplifier will saturate very rapidly! In order to keep the output voltage from saturating, it is necessary to prevent the input signal from rising too high. If the maximum output voltage allowable is 10 volts, then the maximum input signal is 10 volts/10 000 or 1 millivolt! Obviously, it is necessary to keep the RC time constant within certain bounds.

How to solve the problem. Fortunately, there are some design tactics that allow keeping the integration aspects of the circuit, while removing the problems. A practical integrator is shown in Fig. 2-15. The heart of this circuit is a BiMOS operational amplifier, type CA-3140, or an equivalent BiFET device. The reason why this works so well is that it has a low input bias current (being a MOSFET input design).

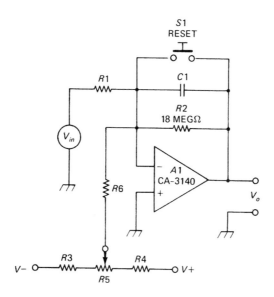

FIGURE 2-15 Practical integrator circuit.

Capacitor $C1$ and resistor $R1$ in Fig. 2-15 form the integration elements, and are used in the transfer equation to calculate performance. Resistor $R2$ is used to discharge $C1$ to prevent DC offsets on the input signal and the op-amp itself from saturating the circuit. The RESET switch is used to set the capacitor voltage back to zero (to prevent a K factor offset) before the circuit is used. In some measurement applications the circuit is initialized by closing S1 momentarily. In actual circuits, S1 may be a mechanical switch, an electromechanical relay, a solid-state relay, or a CMOS electronic switch.

If there is still a minor drift problem in the circuit, then potentiometer $R5$ can be added to the circuit to cancel it. This component adds a slight counter-current to the inverting input through resistor $R6$. To adjust this circuit, set $R5$ initially to mid-range. The potentiometer is adjusted by shorting the V_{in} input to ground (setting $V_{in} = 0$), and then measuring the output voltage. Press S1 to discharge $C1$, and note the output voltage (it should go to zero). If V_o does not go to zero, then turn $R5$ in the direction that counters the change of V_o. This change can be observed after each time RESET switch S1 is pressed. Keep pressing S1 and then making small changes in $R5$, until the setting is found at which the output voltage stays very nearly zero, and constant, after S1 is pressed (there may be some very long-term drift).

Figure 2-16 shows the practical version of the differentiator circuit. The differentiation elements are $R1$ and $C1$, and the previous equation for the output voltage is used. Capacitor $C2$ has a small value (1 pF to 100 pF), and is used to alter the frequency response of the circuit in order to prevent oscillation or ringing on fast risetime input signals. Similarly, a 'snubber'

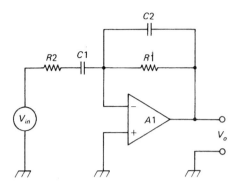

FIGURE 2-16 Practical differentiator circuit.

resistor (R2) in the input also limits this problem. The operational amplifier can be almost any type with a fast enough slew rate, and the CA-3140 is often recommended. The values of R2 and C2 are often determined by rule-of-thumb, but their justification is taken from the Bode plot of the circuit.

2-7 LOGARITHMIC AND ANTI-LOG CIRCUITS

Logarithmic amplifiers are often used in instrumentation circuits, especially where data compression is required. The overall transfer equation for an operational amplifier circuit is determined by the transfer equation of the feedback network. As might be guessed from this fact, a logarithmic transfer equation can be created in an operational amplifier circuit by placing a nonlinear element in the negative feedback loop. An ordinary PN junction transistor meets the requirement. A logarithmic amplifier is one that has a transfer equation of the form of either:

$$V_o = k \ln(V_{in}) \qquad (2\text{-}17)$$

or

$$V_o = k \log(V_{in}) \qquad (2\text{-}18)$$

Figure 2-17 shows the basic circuit for an inverting logarithmic amplifier. As with any inverting amplifier, we can assume that the summing junction potential is zero because the noninverting input is grounded.

In basic transistor theory the base–emitter voltage of the transistor is given by:

$$V_{b-e} = \frac{KT}{q} \ln\left[\frac{I_c}{I_s}\right] \qquad (2\text{-}19)$$

where:
V_{b-e} is the base–emitter potential in volts (V)
K is Boltzmann's constant (1.38×10^{-23} J/K)

52 SIGNAL PROCESSING CIRCUITS

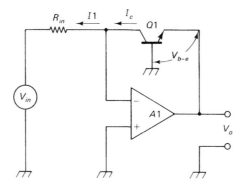

FIGURE 2-17 Logarithmic amplifier circuit.

T is the temperature in kelvin (K)
q is the electronic charge (1.6×10^{-19} coulombs)
ln indicates the natural (or 'base-e') logarithms
I_c is the collector current of the transistor in amperes
I_s is the reverse saturation current of the transistor (approximately 10^{-13} A at 300K)

Because the configuration of Fig. 2-17 makes $V_{b-e} = V_o$:

$$V_o = \frac{KT}{q} \ln\left[\frac{I_c}{I_s}\right] \qquad (2\text{-}20)$$

or, for those who prefer base-10 logarithms:

$$\frac{\log X}{\ln X} = 0.4343 \qquad (2\text{-}21)$$

and,

$$\frac{\ln X}{\log X} = 2.3 \qquad (2\text{-}22)$$

so,

$$V_o = \frac{2.3\, KT}{q} \log\left[\frac{I_c}{I_s}\right] \qquad (2\text{-}23)$$

When the constants KT/q are accounted for:

$$V_o = 60 \log\left[\frac{I_c}{I_s}\right] \quad \text{mV} \qquad (2\text{-}24)$$

The equation above demonstrates that the output voltage V_o is a logarithmic function. From KCL it is known that

$$I1 = -I_c \qquad (2\text{-}25)$$

and from Ohm's law:

$$I1 = \frac{V_{in}}{R_{in}} \tag{2-26}$$

Substituting Eqs (2-25) and (2-26) into Eq. (2-24) yields:

$$V_o = 60 \log \left[\frac{I1}{I_s} \right] \quad \text{mV} \tag{2-27}$$

or

$$V_o = 60 \log \left[\frac{V_{in}/R_{in}}{I_s} \right] \quad \text{mV} \tag{2-28}$$

Thus, the output voltage V_o is proportional to the logarithm of the input voltage V_{in}.

The simple circuit of Fig. 2-17 is the one usually published in textbooks, but in a practical sense it only works some of the time due to the realities of non-ideal operational amplifiers. Figure 2-18 shows one common problem: *oscillation*.

The oscilloscope waveform shown in Fig. 2-18 was taken from a 741 op-amp connected in the simple logarithmic amplifier configuration of Fig. 2-17. The input signal was a linear ramp sawtooth signal, so one would expect a logarithmically decreasing output voltage. Unfortunately, the amplifier oscillates in the basic configuration.

A modified version of the circuit is shown in Fig. 2-19. In this circuit a compensation network (R3/R4/C1) is added to prevent the oscillation. The values of the network components (except R4) is found from empirical data based on the following approximations:

$$R3 = \frac{V_{o(max)} - 0.7}{(V_{in(max)}/R1) + (V_{o(max)}/R4)} \tag{2-29}$$

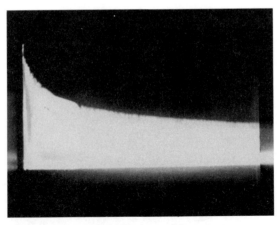

FIGURE 2-18 Oscillation shown at logarithmic amplifier output.

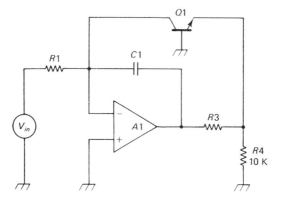

FIGURE 2-19 Compensation to prevent oscillations in logarithmic amplifier.

and

$$C1 = \frac{1}{\pi f R3} \quad (2\text{-}30)$$

Figure 2-20 shows the output signal of Fig. 2-19 with the same linear sawtooth ramp input signal as before. Note that the output signal is clean and free of oscillation, and has the characteristic decreasing logarithmic shape expected of an inverting log amplifier.

Another problem of the logarithmic amplifier is *temperature sensitivity*. Recall that the operant equation for the logarithmic amplifier is:

$$V_o = \frac{KT}{q} \ln\left[\frac{V_{in}/R_{in}}{I_s}\right] \quad (2\text{-}31)$$

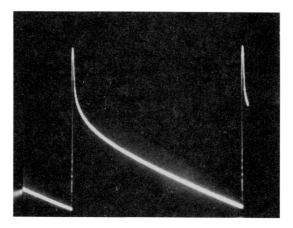

FIGURE 2-20 Cleaned up output waveform of logarithmic amplifier.

SIGNAL PROCESSING CIRCUITS

The T term in the equation is temperature in kelvin (note: room temperature is approximately 300K). Temperature is a variable rather than a constant, so we can expect the output voltage to be a function of both the applied input signal voltage and the temperature of the b-e junction in the transistor. This temperature is, in turn, a function of ambient temperature. In order to prevent pollution of the output signal data it is necessary to *temperature compensate* the logarithmic amplifier circuit. Figure 2-21 shows two approaches to the temperature compensation job. The value of $R1$ in Fig. 2-21A is approximately 15.7 times the temperature of the thermistor (R_t) at room temperature.

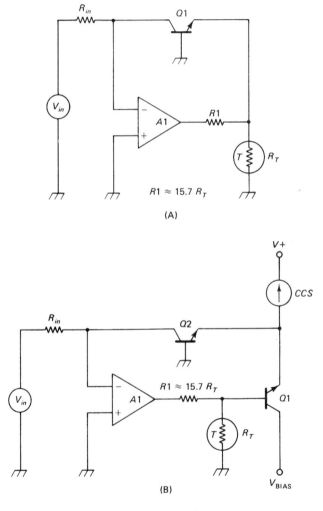

FIGURE 2-21 Two approaches to temperature compensation; (A) simple circuit; (B) active circuit.

56 SIGNAL PROCESSING CIRCUITS

The circuit of Fig. 2-21B is a little more complex, but also offers greater dynamic range than the previous circuit.

2-7.1 Anti-log amplifiers

The *anti-log amplifier* performs the inverse function of the logarithmic amplifier. The output voltage from the anti-log amplifier is:

$$V_o = K \text{ antilog}(V_{in}) \qquad (2\text{-}32)$$

The simplified circuit for a conventional anti-log amplifier is shown in Fig. 2-22. Note that, again, a PN junction from a bipolar transistor is used because of its logarithmic transfer function. But note here that the respective positions of the transistor and resistor are reversed from the logarithmic amplifier.

From basic operational amplifier theory and Kirchhoff's current law (KCL):

$$|I1| = |I2| \qquad (2\text{-}33)$$

Further, by Ohm's law and the fact that the summing junction (point A) is at virtual ground (zero) potential:

$$I2 = \frac{V_o}{R1} \qquad (2\text{-}34)$$

In Fig. 2-22 voltage V_{in} is applied across the b–e junction of $Q1$, so $V_{b-e} = V_{in}$. Current $I1$ is the collector current of $Q1$, so $I_c = I1$. Recall Eq. (2-19):

$$V_{b-e} = \frac{KT}{q} \ln\left[\frac{I_c}{I_s}\right] \qquad (2\text{-}35)$$

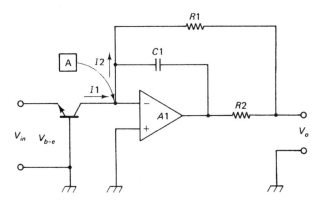

FIGURE 2-22 Anti-logarithmic amplifier.

which can be rewritten to the form:

$$V_{in} = \frac{KT}{q} \ln\left[\frac{I1}{I_s}\right] \tag{2-36}$$

Because $I1 = I2$,

$$V_{b-e} = \frac{KT}{q} \ln\left[\frac{V_o/R1}{I_s}\right] \tag{2-37}$$

Collecting terms:

$$\frac{qV_{in}}{KT} = \ln\left[\frac{V_o/R1}{I_s}\right] \tag{2-38}$$

Taking the anti-log of Eq. (2-38):

$$e^{qV_{in}/KT} = \frac{V_o R1}{I_s} \tag{2-39}$$

$$V_o = \frac{I_s e^{qV_{in}/KT}}{R1} \tag{2-40}$$

Equation (2-40) is the transfer equation for the anti-log amplifier.

2-7.2 Special function circuits

Circuit elements can be combined into a single circuit to form a special function circuit. The combined circuit can be fabricated either in the monolithic IC form, or as a hybrid. Although the range of possible special function circuits is nearly limitless, a few examples serve to illustrate the concept.

Multi-function converter. Figure 2-23 shows the block diagram for a Burr-Brown 4301 or 4302 Multi-Function Converter device. This circuit can perform analog multiplication or division, plus serve such functions as squaring, square rooting, taking other roots, exponentiation, sine, cosine, arctan and root sum squares (RSS).

The 4301/4302 devices consist of a logarithmic amplifier, a log-ratio amplifier, a summer and an anti-log amplifier (see Fig. 2-23). The transfer function for this circuit is:

$$V_o = V_y \left[\frac{V_Z}{V_X}\right]^m \tag{2-41}$$

The output signal can swing to +10 volts output, and will supply 5 mA of output current. The three input signal voltages (V_X, V_Y and V_Z) must be ±10 volts.

The function performed by these devices depends on the nature of the resistor network connected between pins 6, 11 and 12. Figure 2-24 shows the three types of resistor network used to program the 4301/4302. The exponent m (see Eq. (2-41)) determines the function performed. The value of m is set

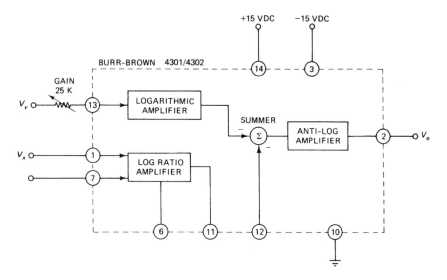

FIGURE 2-23 Burr-Brown 4301 multi-purpose function module.

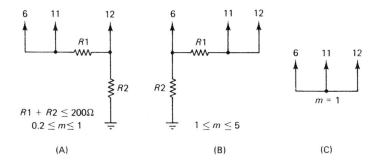

FIGURE 2-24 Gain setting networks for 4301 circuits.

by the external resistors, and can be any value between 0.2 and 5. For root functions $m < 1$. For example, to perform the square root function $m = 0.5$. The resistor network that yields this function is shown in Fig. 2-24A. The values of the resistors for this type of function is set by:

for $m < 1$:

$$m = \frac{R2}{R1 + R2} \tag{2-42}$$

Values greater than one ($1 < m < 5$) allow the 4301/4302 devices to be used as squarers and exponentiators. For these functions use the resistor network in Fig. 2-24B. The values of the resistors are found from:

$$m = \frac{R1 + R2}{R2} \tag{2-43}$$

Using the 4301/4302 as an analog multiplier or divider requires no resistor networks at all. In these circuits $m = 1$, and all three terminals are strapped together (Fig. 2-24C). In this application, two inputs will represent signals, while the third represents a constant and is set to a fixed voltage. For example, to build an analog multiplier V_Y and V_Z are used as signal inputs, while V_X is connected through a resistor to a fixed reference potential. The transfer function in that case is:

$$V_o = K V_Y V_Z \tag{2-44}$$

where K is proportional to V_X.

Similarly, to make an analog divider, V_Y becomes the constant (K) and the input signals are V_X and V_Z. The transfer equation is:

$$V_o = K \frac{V_Z}{V_X} \tag{2-45}$$

Devices such as the Burr-Brown 4301/4302 are usable for a wide variety of analog signal processing tasks.

Programmable gain amplifiers. Amplifier gains must sometimes be changed for different applications. In other situations the gain must be changed in an effort to calibrate a system. A programmable gain amplifier allows the gain to be set (or 'programmed') from an external source. In some simple cases the programming is done by setting a bias current. A single resistor is connected between the programming terminal and a reference DC power supply. This form of amplifier does not allow the type of control that is amenable to either computer control or control by other digitally oriented circuits.

There are at least two methods for digitally programming an amplifier. One is to use a multiplying digital-to-analog converter (MDAC). All DACs provide an output that is proportional to a binary word and a DC reference input. In a multiplying DAC, or MDAC, the DC reference is supplied from an external source. It is possible on some MDACs to use either a fixed DC source, a slowly varying nearly-DC signal, an AC signal with a DC offset, or a symmetrical AC signal (in which case the offset is provided by the designer) as the reference. If the 'reference' is the signal to be amplified, then the output will be a function of the reference signal input and the binary word applied to the digital inputs.

Another approach to designing the programmable amplifier is shown in Fig. 2-25. The amplifier circuit is the three-device instrumentation amplifier

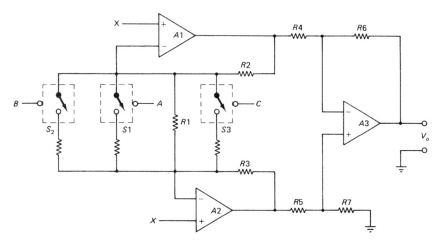

FIGURE 2-25 Programmable gain instrumentation amplifier.

circuit. Gain is set by:

$$A_v = \left[\frac{2R2}{R1} + 1\right]\left[\frac{R6}{R4}\right] \qquad (2\text{-}46)$$

Assuming $R2 = R3$, $R4 = R5$ and $R6 = R7$.

By making all resistors fixed, the programmable amplifier gain can be controlled solely by varying $R1$. In Fig. 2-25 this function is performed by shunting additional resistors across a relatively high value of fixed resistor that is always in the circuit. The switches ($S1$ through $S3$) are CMOS electronic switches with active-LOW digital control terminals. When the control terminals (A, B and C) are HIGH, then the switch is open, and when the control terminal is LOW the switch is closed. Gain is set by selecting any (or several) switches to be closed in order to alter the resistance of $R1$ as seen by the amplifier.

The circuit of Fig. 2-25 is actually quite crude compared with programmable amplifiers offered by device manufacturers. In some of those products more than one resistance parameter can be varied, resulting in gain settings of small increments from unity to 1024.

Thermocouple amplifiers. A *thermocouple* is a junction formed of two dissimilar metals. These devices are used to measure temperature over a very wide range (>2700°C), and can be very precise when properly built and calibrated. The thermocouple works because of the *Seebeck effect*. All metals have a certain work function that defines how much energy is needed to loosen electrons from the atom. If two metals with different work functions are joined together in a fused junction, then the difference between work functions, a voltage, is generated across the two ends of the wires.

SIGNAL PROCESSING CIRCUITS

TEMPERATURE MEASUREMENT COMPONENTS
Temperature Transducer Signal Conditioners

AD594/AD595

Pretrimmed for Type J (AD594) or Type K (AD595) Thermocouples
Can Be Used with Type T Thermocouple Inputs
Low Impedance Voltage Output: 10mV/°C
Built-In Ice Point Compensation
Wide Power Supply Range: +5V to ±15V
Low Power: <1mW typical
Thermocouple Failure Alarm
Laser Wafer-Trimmed to 1°C Calibration Accuracy
Set-Point Mode Operation
Self-Contained Celsius Thermometer Operation
High Impedance Differential Input
Side-Brazed DIP or Low Cost CERDIP

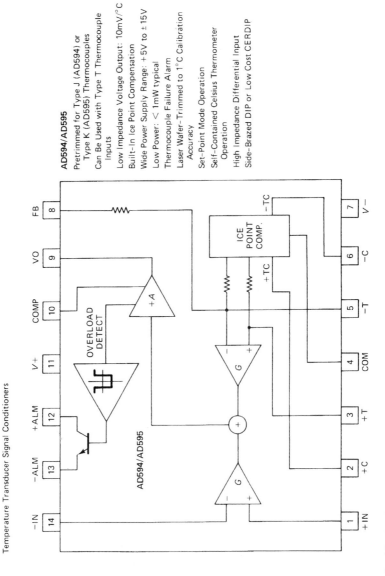

FIGURE 2-26 AD-594 thermocouple amplifier.

That voltage is proportional to the temperature of the junction and the work function differential. Thermocouples are given letter designations (e.g. K and T) to denote both a certain characteristic and a pair of specific metals used to form the junction.

The inverse function, called the *Peltier effect*, is responsible for the so-called 'solid-state' refrigerators; applying a voltage across the loose ends of the thermocouple causes one wire to lose heat and the other to absorb heat.

It is common practice in thermocouple temperature measurements to use two thermocouples. One is placed in a zero degree Celcius ice bath and becomes a reference pole. Iced water is exactly $0°C$ when there is equilibrium between ice and liquid water in the bath. The other thermocouple is the measurement pole, and is connected in series with the ice-point thermocouple. A differential amplifier measures the difference potential between the two thermocouples, and the resulting output voltage is (within the linearity of the system) proportional to the temperature difference between $0°C$ and the measured temperature. This arrangement is called 'ice point compensation'.

Analog Devices, Inc. makes a special function IC device that is used as a transducer signal conditioner. A DC differential input amplifier converts the thermocouple signal into a single-ended output signal, V_o. The AD-594 and AD-595 devices (Fig. 2-26) contain all of the circuitry necessary to properly operate the thermocouple. These devices also contain electronic circuitry that performs the ice-point compensation function mentioned above. Both Type-T and Type-K thermocouples can be accommodated. The output of the IC will produce a scaling function of 10 mV/$°C$.

2-8 SUMMARY

1. Analog circuits are often used for signals processing functions including *amplification, integration, differentiation, exponentiation* and *logarithmic amplification*.
2. *Chopper amplifiers* use electronic switching at the input and output terminals of an AC amplifier to process slowly changing or DC signals. The principal advantages of the chopper amplifier are drift and noise reduction.
3. *Carrier amplifiers* are used to process transducer signals. The excitation signal is an AC carrier between 1 kHz and 25 kHz. The signal is processed in an AC amplifier, and the analog data is recovered in a phase sensitive detector.
4. *Lock-in amplifiers* are used to process very low-level signals. Two forms are found. One is a very narrow band, high-Q, AC carrier amplifier. The other is an autocorrelation amplifier.
5. *Integrators* find the area underneath a voltage versus time function over a specified interval. *Differentiators* find the instantaneous rate of change of an analog signal. Both circuits can be made using the inverting follower configuration. In the case of the integrator an input resistor and a feedback capacitor are used. In the case of the differentiator an input capacitor and a feedback resistor are used.

6. In a *logarithmic amplifier* a nonlinear PN junction is used in the feedback network of an inverting follower to provide a logarithmic transfer function. In the *anti-log amplifier* the transistor is connected in series with the input signal of the inverting follower.

7. The various forms of analog signals processing circuit are often combined into a single package to form *special function circuits*.

2-9 STUDENT EXERCISES

1. Select several resistors in the 10 kohms to 1 megohm range, and several capacitors in the 0.001 µF to 10 µF range. Fashion an *RC* integrator (Fig. 2-12A) from one resistor and one capacitor. Apply the output of a function generator to the *RC* integrator, and observe both input and output waveforms on a dual-beam oscilloscope. Vary both the frequency of the signal source and the output waveform to see how the *RC* integrator affects the signal.

2. Perform the exercise above, but connect the components as a differentiator (Fig. 2-12B).

3. Build a Miller integrator circuit such as Fig. 2-14 using an input resistor of 100 kohms and a feedback capacitor of 0.1 µF. Apply a squarewave of $1/RC$ Hz to the input and observe the output signal on an oscilloscope. Vary the frequency of the signal to see how the circuit affects the waveform. Explain any anomalies observed.

4. Build a Miller integrator circuit such as Fig. 2-15, and compare the operation to that of the previous exercise.

5. Design and build a logarithmic amplifier such as Fig. 2-19. First use a 1 volt sawtooth to observe the operation, and then other waveforms.

2-10 RECAPITULATION

Now return to the objectives and Pre-quiz questions at the beginning of the chapter and see how well you can answer them. If you cannot answer certain questions, place a check mark to each and review the appropriate parts of the text. Next, try to answer the questions and work the problems below, using the same procedure.

2-11 QUESTIONS AND PROBLEMS

1. Draw the schematic for a simple chopper amplifier using a mechanical SPDT switch. Similarly, draw a circuit replacing the mechanical switch with CMOS electronic switches.

2. An operational amplifier in the noninverting configuration has a noise figure of $75 \text{ nV}/(\sqrt{\text{Hz}})$. Calculate the output RMS noise when the bandwidth is 2000 Hz.

3. Draw the schematic for the input circuit of a typical differential chopper amplifier.

4. Draw the block diagram for a carrier amplifier, and explain the function of the various stages.
5. What is the advantage of a phase sensitive detector (PSD).
6. Draw the block diagram for a lock-in amplifier.
7. (a) Draw the circuit diagram for a Miller integrator; (b) Write the transfer function for the Miller integrator; (c) What are alternate names for the Miller integrator.
8. Draw the circuit diagram for an operational amplifier differentiator.
9. The time constant for a differentiator should be _____ relative to the period of the input signal; the time constant for an integrator circuit should be _____ relative to the period of the input signal.
10. How is an integrator used to make a quadrature circuit, i.e. a circuit that has both sine and cosine outputs.
11. Calculate the gains for Miller integrator circuits in which the following RC time constants are used: 1 megohm and 0.01 µF, 10 kohms and 0.1 µF, 100 kohms and 0.001 µF.
12. A Miller integrator consists of a 100 kohm input resistor and a 1 µF feedback capacitor. The maximum allowable output potential is +11 volts. Calculate the time required to rise from an initial condition of 0 volts to saturation when a 0.1 Vdc signal is applied to the input.
13. Draw the circuit for: (a) logarithmic amplifier; (b) anti-log amplifier.
14. An electronic integrator is used on a medical blood pressure monitor to find the _____ _____ of the applied analog pressure waveform. This signal is the area under the _____ curve.
15. What are the advantages of: (a) a carrier amplifier, (b) a chopper amplifier?
16. A pH (acid-base measurement) transducer has a drift factor of 1.4 mV/°C, and the gain-of-10 voltage amplifier processing the signal has a drift of 0.5 mV/°C. Calculate the total drift component if the amplifier ambient temperature rises from 30°C to 45°C, and the solution in which the transducer is immersed rises from +25°C to +38°C.
17. A 400 Hz sinewave is applied to the input of a Miller integrator circuit. Sketch the output waveform expected from this circuit.
18. Derive the transfer equation of the operational amplifier Miller integrator using the Kirchhoff's current law method.
19. Derive the transfer equation of an operational amplifier differentiator using the Kirchhoff's current law method.
20. Derive the transfer equation of the simple op-amp logarithmic amplifier.
21. Describe some of the problems that might be encountered in designing and building practical operational amplifier Miller integrators, and some possible solutions to those problems.
22. Describe some of the problems that might be encountered in designing and building practical operational amplifier differentiators, and some possible solutions to those problems.

23. Design a practical Miller integrator circuit in which the output voltage will rise at a rate of 1 volt per second when a 100 mV DC input signal is applied.
24. Describe why the logarithmic amplifier must be temperature compensated in order to avoid a large temperature sensitive error. How can the amplifier be temperature compensated? Draw a sketch of the circuit.
25. A digital-to-analog converter that can be used as a programmable gain amplifier is also called a _____ DAC.
26. Describe the properties of, and differences between, analog signals, sampled analog signals, and digital signals.
27. Describe how a sampled analog signal might be integrated. Use a sketch if appropriate.
28. What is the purpose of resistor $R2$ in Fig. 2-15? And the purpose of $S1$?
29. Why is the resistor network $R3$ through $R6$ required on practical integrators such as Fig. 2-15, even though it is not always shown on textbook integrator circuits?
30. Describe the functions of resistor $R2$ and capacitor $C2$ in Fig. 2-16.
31. Describe the functions of $C1/R3$ in Fig. 2-19.
32. Calculate the values for $R3$ and $C1$ in Fig. 2-19 if the maximum operating frequency is 500 Hz, $R1$ is 12 kohms, the maximum input signal is +5 volts, and the maximum output voltage is +16 volts.
33. Design a programmable gain amplifier (PGA) such as Fig. 2-25 with following differential voltage gains: ×100, ×500, and ×1000.

CHAPTER 3

Measurement and instrumentation circuits

OBJECTIVES
1. Learn the basics of instrumentation circuits using linear integrated circuit devices.
2. Understand instrument signal sources such as sensors, transducers, electrodes and so forth.
3. Know the parameters affecting instrument circuit designs.
4. Be able to design simple electronic instruments.

3-1 PRE-QUIZ

These questions test your prior knowledge of the material in this chapter. Try answering them before you read the chapter. Look for the answers (especially those you answered incorrectly) as you read the text. After you have finished studying the chapter try answering these questions again, and those at the end of the chapter (see Section 3-7).

1. A strain gage Wheatstone bridge transducer has a sensitivity factor of 35 μV/V/Torr of pressure, and is excited by a 10.00 volt DC source. Calculate the output voltage if a 420 Torr pressure is applied.
2. Using the parameters of the problem above calculate the gain of a DC differential amplifier required to produce an output sensitivity of 10 mV/Torr, i.e. an output potential of 4.200 volts when the applied pressure is 420 Torr.
3. Define 'transduction'.
4. _____ metric measurement is often used when the excitation potential or some other parameter of the measurement system is likely to change independently of the measured parameter.

3-2 INTRODUCTION

One of the functions of this book is to give the student familiarity with linear integrated circuits *and their applications*. Electronic instrumentation for control and measurement systems is one category of such applications. You will study broad categories of signal sources (transducers and electrodes) and the circuitry needed to process these signals.

3-3 TRANSDUCTION AND TRANSDUCERS

Transduction is the process of *changing energy from one form to another for purposes of measurement, tabulation or control.* Transducers and sensors, are the eyes and ears of the electronic instrument. These devices convert assorted forms of energy from physical systems (e.g. temperature, pressure, etc.) into electrical energy in a form usable in electronic systems. For example, a pressure transducer might convert fluid pressure to an analogous voltage or current.

In this section the most basic forms of transducer are discussed. The intent is not to provide a complete catalog of available types, but rather to introduce generic classes of transducers. The text material begins with a discussion of *piezoresistive strain gages*.

There are many different forms of transducer that use resistive strain gage elements, and most of them are based on the Wheatstone bridge circuit. Various physical parameters can be measured with strain gage transducers, including force, displacement, vibration, and both liquid and gas pressure. For example, if a hospital patient requires intensive care, then the physician may order continuous electronic blood pressure monitoring through an indwelling fluid-filled catheter inserted into an artery. The transducer used to measure blood pressure may be a *Wheatstone bridge strain gage* (although other types exist). This section discusses how such strain gages work, and how to make them operate properly in practical cases.

3-3.1 Piezoresistivity

All electrical conductors possess electrical resistance. The resistance of any specific conductor is directly proportional to its length (see Fig. 3-1), and inversely proportional to its cross-sectional area. Resistance is also directly proportional to a physical property of the conductor material called *resistivity*. The relationship among *length* (L), *cross-sectional area* (A), and *resistivity* (ρ) is:

$$R = \frac{\rho L}{A} \tag{3-1}$$

Piezoresistivity is a phenomena in which the resistance changes when either length or cross-sectional area are changed. That is, electrical resistance

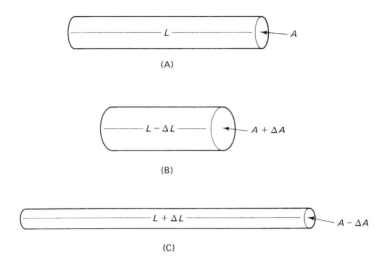

FIGURE 3-1 (A) Cylindrical conductor at rest; (B) in compression; (C) in tension.

changes in response to mechanical deformation. Figure 3-1A shows a cylindrical conductor with a resting length L_o and a cross-sectional area A_o. When a *compression force* is applied, as in Fig. 3-1B, the length decreases and the cross-sectional area increases. This situation results in a decrease in the electrical resistance.

Similarly, when a *tension force* is applied (Fig. 3-1C) the length increases and the cross-sectional area decreases, so the electrical resistance increases. Provided that the physical change is small, and the conductor's elastic limit is not passed, the change of electrical resistance is a nearly linear function of the applied force, so can be used to make measurements of that force.

Strain gages. A *strain gage* is a piezoresistive element made of either wire, metal foil or a semiconductor material, and is designed to create a *resistance change when a force is applied*. Strain gages can be classified as either *bonded* or *unbonded* types. Figure 3-2 shows both methods of construction.

The *unbonded strain gage* is shown in Fig. 3-2A. It consists of a wire resistance element stretched taut between two flexible supports. These supports are configured in a manner that places either a tension or compression force on the taut wire when an external force is applied. In the particular example shown, the supports are mounted on a thin metal diaphragm that flexes when a force is applied. Force $F1$ will cause the flexible supports to spread apart, placing a tension force on the wire and increasing its resistance. Alternatively, when force $F2$ is applied, the ends of the supports tend to move closer together, effectively placing a compression force on the wire element and thereby reducing its resistance. The resting

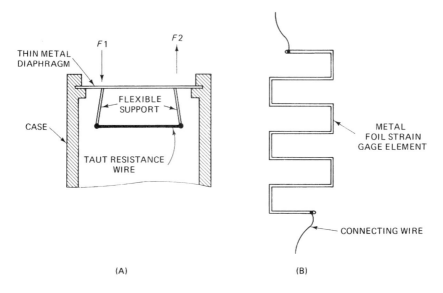

FIGURE 3-2 (A) Unbonded strain gage element; (B) bonded strain gage.

condition is slight tautness, which implies a small normal tension. Force $F1$ increases the normal tension, and force $F2$ decreases the normal tension.

The *bonded strain gage* is shown in Fig. 3-2B. In this type of strain gage a wire, foil or semiconductor element is cemented directly to a thin metal diaphragm. When the diaphragm is flexed, the element deforms to produce a resistance change.

The linearity of both types of strain gage can be quite good, provided that the elastic limits of the diaphragm and element are not exceeded. A metallic diaphragm can be distended linearly and will regain its shape only as long as the applied force is less than the force required to exceed the limit of elasticity. It is therefore necessary to ensure that the change of length is only a small percentage of the resting length.

In the past, bonded strain gages were more rugged, but less linear than unbonded models. Although this may have been the situation at one time, recent technology has produced highly linear, reliable units of both types of construction.

3-3.2 The Wheatstone bridge

The Wheatstone bridge is a 19th-century circuit that still finds widespread application in modern electronic instrument circuits. The classic form of Wheatstone bridge is shown in Fig. 3-3. There are four resistive arms in the bridge, which are labeled $R1$, $R2$, $R3$ and $R4$. The excitation voltage (V) is applied across two of the nodes, while the output signal is taken from the alternate two nodes (labeled C and D). This circuit can be modeled as two

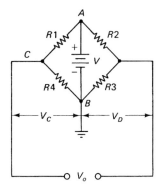

FIGURE 3-3 Wheatstone bridge circuit.

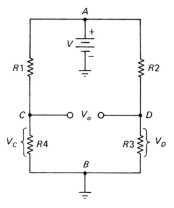

FIGURE 3-4 Strain gage circuit redrawn as two voltage divider networks.

series voltage dividers in parallel, one consisting of $R1$ and $R4$ and the other by $R2$ and $R3$ (See Fig. 3-4).

The output voltage from a Wheatstone bridge is the difference between the voltages at points C and D. The mathematics of the circuit reveals that the output voltage will be zero when the ratio $R4/R1$ is equal to the ratio $R3/R2$. If these ratios are not kept equal, as is the case when one or more elements is a strain gage not at rest, then an output voltage is produced that is proportional to both the applied excitation voltage and the change of resistance.

Strain gage circuitry. Before the resistive strain gage can be useful it must be connected into a circuit that will convert its resistance changes into a current or voltage output that is proportional to the applied force. Most strain gage applications use voltage-output circuits.

Figure 3-5 shows several popular forms of circuit. The circuit in Fig. 3-5A is both the simplest and least useful (although not useless!); it is sometimes

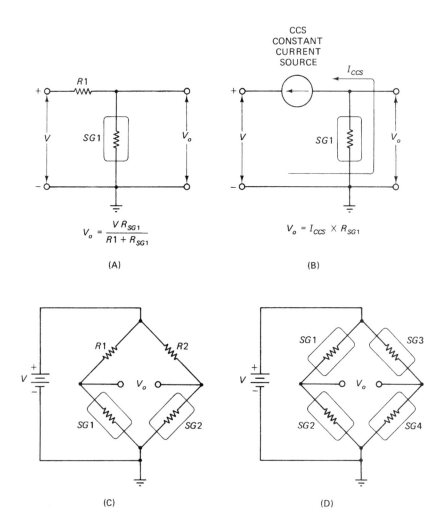

FIGURE 3-5 (A) Single strain gage half-bridge; (B) single strain gage circuit with constant current source drive; (C) two strain gage Wheatstone bridge; (D) four strain gage Wheatstone bridge.

called the *half-bridge* circuit, or *voltage divider* circuit. The strain gage element of resistance R is placed in series with a fixed resistor, $R1$, across a stable DC voltage, V. The output voltage V_o is found from the voltage divider equation:

$$V_o = \frac{VR}{R+R1} \tag{3-2}$$

Equation (3-2) describes the output voltage V_o when the transducer is at rest (i.e. nothing is stimulating the resistive element). When the element is stimulated, however, its resistance changes a small amount, h. The output voltage in that case is:

$$V_o = \frac{V(R+h)}{(R \pm h) + R1} \qquad (3\text{-}3)$$

Another form of half-bridge circuit is shown in Fig. 3-5B. The strain gage is connected in series with a constant current source (CCS), which will maintain current I at a constant level regardless of changes in the strain gage resistance. In this case, $V_o = I(R \pm h)$.

Both of the half-bridge circuits suffer from one major defect: an output voltage V_o will always be present regardless of the value of the stimulus applied to the transducer. Ideally, in any transducer system the output voltage should be zero when the applied stimulus is zero. For example, when a gas pressure transducer is open to atmosphere, the gage pressure is zero so the transducer output voltage should ideally also be zero. Secondly, the output voltage should be proportional to the value of the stimulus when the stimulus is not zero. A Wheatstone bridge circuit can have these properties if properly designed. Strain gage elements can be used for one, two, three or all four arms of the Wheatstone bridge.

Figure 3-5C shows a circuit in which two strain gages ($SG1$ and $SG2$) are used in two arms of a Wheatstone bridge, with fixed resistors $R1$ and $R2$ forming the alternate arms of the bridge. It is usually the case that $SG1$ and $SG2$ are configured so that their actions oppose each other; that is, under any given stimulus, element $SG1$ will assume a resistance $R+h$, and $SG2$ will assume a resistance $R-h$, or vice versa. If both resistances changed in the same direction, then bridge balance is maintained and no output voltage is generated.

One of the most linear forms of transducer bridge is the circuit of Figure 3-5D in which all four bridge arms contain strain gage elements. In most such transducers all four strain gage elements have the same resistance (R), which will usually be a value between 50 and 1000 ohms.

Recall that the output from a Wheatstone bridge is the difference between the voltages across the two half-bridges. It is possible to calculate the output voltage for any of the standard configurations from the equations given below. Assuming that all four bridge arms nominally have the same resistance, the output is:

One active element

$$V_o = \frac{Vh}{4R} \qquad (3\text{-}4)$$

(accurate to $\approx \pm 5\%$, provided that $h < 0.1$).

Two active elements

$$V_o = \frac{Vh}{2R} \qquad (3\text{-}5)$$

Four active elements

$$V_o = \frac{Vh}{R} \tag{3-6}$$

where:
  V_o is the output potential
  V is the excitation potential
  R is the bridge arm nominal resting resistance
  h is the change of bridge arm resistance (ΔR) under the applied stimulus

An example is provided for a bridge with all four arms active (similar to Fig. 3-5D).

EXAMPLE 3-1

A force transducer is used to measure the weight of small objects. It has a resting resistance of 200 ohms, and an excitation potential of +5 volts DC is applied. When a 1 gram weight is placed on the transducer diaphragm, the resistance of the bridge arms changes by 4.1 ohms. What is the output voltage?

Solution

$$V_o = \frac{Vh}{R}$$

$$V_o = \frac{(5 \text{ volts})(4.1 \, \Omega)}{200 \, \Omega}$$

$$V_o = \frac{20.5}{200} \text{ volts}$$

$$V_o = 0.103 \text{ volts} = 103 \text{ millivolts}$$

Transducer sensitivity (ψ). Most readers will probably work with a known transducer for which the sensitivity is specified. The *sensitivity factor* (ψ) relates the *transducer output voltage* (V_o) to the *applied stimulus value* (Q) and *excitation voltage* (V). In most cases, the transducer manufacturer will specify a number of *microvolts (or millivolts) output potential per volt of excitation potential per unit of applied stimulus*. In other words:

$$\psi = V_o/V/Q \tag{3-7}$$

or, written another way:

$$\psi = \frac{V_o}{V \times Q} \tag{3-8}$$

where:
  ψ is the sensitivity factor (μV/V/Q)
  V_o is the output potential (μV)
  V is the excitation potential (V)
  Q is one unit of applied stimulus

If the sensitivity factor is known, then it is possible to calculate the output potential:

$$V_o = \psi V Q \tag{3-9}$$

Equation (3-9) is the one that is most often used in circuit design.

EXAMPLE 3-2

A fluid pressure transducer is used for measuring human and animal blood pressures through an indwelling fluid catheter. It has a sensitivity factor (ψ) of 5 µV/V/Torr (note: 1 Torr = 1 mmHg). Find the output potential when the excitation potential is +7.5 volts DC and the pressure is 400 Torr.

Solution

$$V_o = \psi V Q$$

$$V_o = \frac{5 \text{ µV}}{V \times T} \times (7.5 \text{ volts}) \times (400 \text{ Torr})$$

$$V_o = (5 \times 7.5 \times 400) \text{ µV}$$

$$V_o = 15\,000 \text{ µV} = 15 \text{ mV} = 0.015 \text{ volts} \quad \blacksquare$$

Balancing and calibrating a bridge transducer. Few Wheatstone bridge transducers meet the ideal condition in which all four bridge arms have exactly equal resistances and exactly equal resistance changes per unit of stimulus. In fact, the bridge resistance specified by the manufacturer is only a nominal value, and the actual value may vary quite a bit from the specified value. There will inevitably be an offset voltage (i.e. V_o is not zero when Q is zero). Figure 3-6 shows circuits that will balance the bridge when the stimulus is zero.

The method shown in Fig. 3-6A depends on the fact that the ratio of the two bridge halves must be equal for null to occur [$(R_{SG1}/R_{SG2}) = (R_{SG3}/R_{SG4})$]. If the bridge is unbalanced, then potentiometer $R1$ can be used to trim out the differences by adding a little more resistance to one half of the bridge than the other. The excitation potential V is applied to the wiper of the potentiometer.

The method shown in Fig. 3-6B depends on current injection into one node of the bridge circuit. Potentiometer $R1$ in Fig. 3-6B is usually a precision type with five to fifteen turns required to cover the entire range. The purpose of the potentiometer is to inject a balancing current (I) into the bridge circuit at one of the nodes adjacent to an excitation node. Potentiometer $R1$ is adjusted, with the stimulus at zero, for zero output voltage.

Another application for this type of circuit is injecting an intentional offset potential. For example, on a digital weighing scale such a circuit may be used to adjust for the 'Tare weight' of the scale, which is the sum of the platform and all other weights acting on the transducer when no sample is

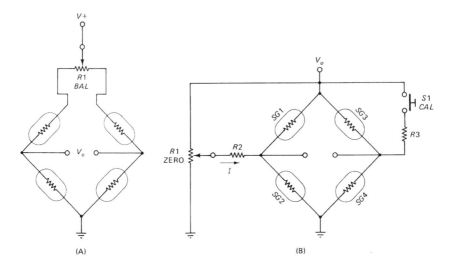

FIGURE 3-6 (A) Balance circuit for Wheatstone bridge; (B) current injection balance circuit for Wheatstone bridge.

being measured on the scale. This is also sometimes called *empty weight compensation.*

Calibration can be accomplished using either of two methods. The most accurate method is to set the transducer up in a system and apply the stimulus. The stimulus is measured by other means and the result is compared with the transducer output. For example, if you are testing a pressure transducer, connect a manometer (pressure measuring device containing a column of mercury) and measure the pressure directly. The result is compared with the transducer output. All transducers should be tested in this manner initially when placed in service and then periodically thereafter.

In less critical applications, however, the alternative (but less accurate) method may be used. It is possible to use a *calibration resistor* to synthesize the offset and thereby allow the electronics to be calibrated. The resistor and CAL switch ($S1$) in Fig. 3-6B is used for this purpose. Resistor $R3$ should have a value of:

$$R3 = \frac{R}{4Q\psi} - \frac{R}{2} \qquad (3\text{-}10)$$

In the equation above we express the output voltage from the sensitivity factor (ψ) as volts instead of microvolts.

Auto-zero circuitry. In the previous section you learned that transducers are rarely perfectly balanced. Even when the arm elements have the same nominal at-rest resistance (R), an offset error may exist due to the fact that the bridge arm resistances each have a certain tolerance, and that causes a difference in the actual values of resistance. Because of this problem, the

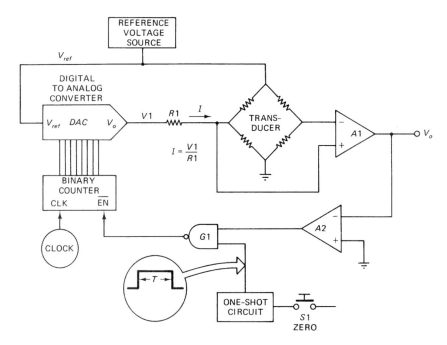

FIGURE 3-7 Auto-zero circuit.

output voltage of the Wheatstone bridge transducer will be non-zero when there is no stimulus applied. The previous section showed a manual means for zeroing or balancing the transducer, so now an *autozero circuit* is presented.

Figure 3-7 shows the block diagram of a typical autozero circuit. The bridge and bridge amplifier ($A1$) are the same as in the other circuits. The offset cancellation current (I) is generated by applying either the voltage output of a *digital-to-analog converter* (DAC) to resistor $R1$, or the output of a current producing DAC to the same point on the bridge as $R1$.

Monitoring the output of the bridge amplifier in Fig. 3-7 is a voltage comparator ($A2$). When V_o is zero, the $A2$ input is zero, so the output is LOW. Alternatively, when the amplifier ($A1$) output voltage is non-zero, the $A2$ output is HIGH. The DAC binary inputs are connected to the digital outputs of a binary counter that is turned on when the *enable* (EN) line is HIGH. Circuit operation is as follows:

1. The operator sets the transducer to zero stimulus (e.g. for pressure transducers the valve to atmosphere is opened).

2. The ZERO button is pressed, thereby triggering the one-shot to produce an output pulse of time T, which makes one input of NAND gate $G1$ HIGH.

3. If voltage V_o is non-zero, then the $A2$ output is also HIGH, so both inputs of $G1$ are HIGH — making the $G1$ output LOW, thereby turning on the binary counter.

MEASUREMENT AND INSTRUMENTATION CIRCUITS   77

4. The binary counter continues to increment in step with the clock pulses, thereby causing the DAC output to rise continuously with each increment. This action forces the bridge output towards null.
5. When V_o reaches zero, the A2 output turns off, stemming the flow of clock pulses to the binary counter, and stopping the action. The DAC output voltage will remain at this voltage level.

Transducer linearization. Transducers are not perfect devices. Although the ideal output function is perfectly linear, real transducers are sometimes highly nonlinear. For Wheatstone bridge strain gages the constraints on linearity include making ΔR (called h in some equations) very small (less than 5%) compared with the at-rest resistance of the bridge arms.

There are several methods for linearization. An analog method is shown in Fig. 3-8A. Here the circuit of the usual single strain gage Wheatstone bridge is modified. The ground end of one bridge arm resistor is lifted, and applied to the output of a null-forcing amplifier, A1. In this case, the resistor element $R(1+h)$ is in the feedback network of operational amplifier A1. Small amounts of nonlinearity are cancelled with this circuit.

Another analog method is shown in Fig. 3-8B. This circuit is based on the Burr-Brown DIV-100 analog divider. Two inputs are provided, D and N. The transfer function of the DIV-100 is:

$$V_o = \frac{10N}{D} \tag{3-11}$$

where:
   V_o is the output signal voltage
   N is the signal voltage applied to the N input
   D is the signal voltage applied to the D input

It can be shown that the bridge output voltage is given by:

$$V_d = \frac{-(V+)h}{2(2+h)} \tag{3-12}$$

If R_n and R_d are the input resistances of the N and D inputs on the DIV-100, respectively, the signal voltages applied to the inputs is found from applying the voltage divider equation:

$$N = \frac{-(V+)hR_n}{(2R1+3R_n)(2+h)} \tag{3-13}$$

and,

$$D = \frac{2(V+)R_d}{(2R1+3R_d)(2+h)} \tag{3-14}$$

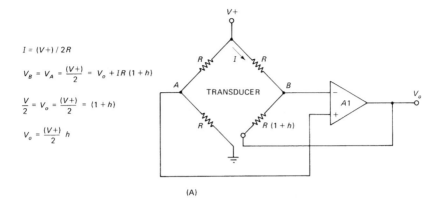

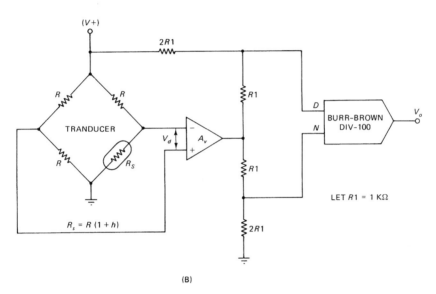

FIGURE 3-8 (A) linearized Wheatstone bridge; (B) using analog divider as a linearizer for bridge circuit.

Substituting Eqs (3-13) and (3-14) into Eq. (3-15), the DIV-100 transfer equation, yields:

$$V_o = \frac{(2R1 + 3R_d)(R_n h)(10)}{(2R1 + 3R_n)(2R_d)} \qquad (3\text{-}15)$$

Collecting terms:

$$V_o = -5h \qquad (3\text{-}16)$$

This circuit is basically a *ratiometric* linearization method because the excitation voltage, $(V+)$, is applied to the DIV-100 inputs through the voltage divider. Thus, the output voltage taken from the DIV-100 is free of variations in the transducer excitation potential... a common form of artifact in other circuits.

For larger degrees of nonlinearity, or for nonlinearity over a large range, other methods must be employed. Figure 3-9 shows a hypothetical transducer transfer function in which a voltage V_o is a function of an applied pressure, P. The ideal transducer will output a linear signal of the form $V_o = mP + b$, where m is the slope of the line and b is the offset.

Before digital computer chips were routinely used in electronic instrumentation applications, special function circuits or *diode breakpoint generators* were used for linearization. In cases where a special function circuit is used the assumption was that the equation of the actual curve is known. The special function circuit generated the inverse of that function and summed it with the input voltage in a difference amplifier. In the case of the diode breakpoint generator an offset voltage is added to or subtracted from the actual input signal to normalize it to the ideal. A reverse biased diode is used to switch on when the input signal voltage exceeds a certain value, causing an offset

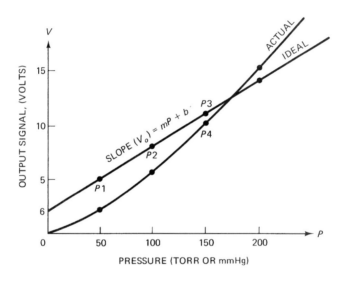

POINT	PRESSURE (TORR)	IDEAL	ACTUAL
P1	50	5	2
P2	100	8	6
P3	150	11	10
P4	200	14	16

FIGURE 3-9 Actual versus ideal calibration curves.

correction bias voltage to be added to the output voltage. This method is rarely used today. It suffered from the temperature sensitivity of the diode switches, and the fact that only a piecewise linear approximation is possible unless a very large number of diode breakpoints was used.

Now that microprocessors are routinely used in even very common instruments, software methods are used to correct transducer error. If the equation defining the actual curve is known, then we could write a software program that cancels the error algebraically. There should exist some polynomial that allows errors to be smoothed out mathematically. Alternatively, the look-up table method of Fig. 3-10 may be used. This example shows only a limited number of data points for simplicity sake, but the actual number would depend upon the bit length of the analog-to-digital (A/D) converter used to convert the analog voltage into a binary word that the computer can digest. An eight-bit A/D converter can represent 2^8 (256_{10}) different values.

The values for the ideal transfer function are stored in a look-up table in computer memory that begins at address location HF000 (the 'H' indicates that hexadecimal or base-16 notation is used). The value HF000 is stored in the X-register. When the A/D binary word is input to the computer it

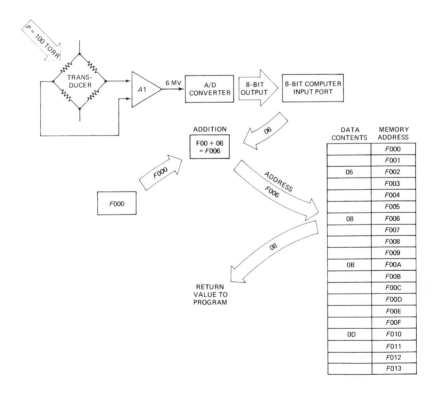

FIGURE 3-10 Software linearization method.

is added to the contents of the X-register. This value becomes the indexed address in the look-up table where the correct value is found. Although a pressure transducer example is shown here, it is useful for almost any form of transducer.

Transducer excitation sources. The DC Wheatstone bridge transducer requires a source of either AC or DC excitation voltage, with DC being the most common. Most transducers require an excitation voltage of 10 volts DC or less. The maximum allowable voltage is critical, and exceeding it will shorten the life expectancy of the transducer. An optimum voltage may be specified for the transducer. This potential is the *highest excitation potential that will not cause thermal drift due to self-heating of the transducer resistance elements*. A typical fluid pressure transducer requires +7.5 volts DC, and operates best (with the least thermal drift) at +5 volts DC. A source of DC must be provided that is stable, within specifications and accurate.

The simplest form of transducer excitation is the zener diode circuit in Fig. 3-11A. A zener diode regulates the voltage sufficiently well for many applications. There are two problems with this circuit, however. First is the fact that the zener potential is typically not a nice reasonable value like 5.0 volts, but will have a value such as 4.7, 5.6, 6.2 or 6.8 volts. The second and most serious defect is *thermal drift* of the zener point. The zener voltage will vary somewhat with temperature in all but certain temperature compensated reference grade zener diodes or band-gap diodes. Unless the application is not critical, or the diode can be kept at a constant temperature, the method of Fig. 3-11A is not generally suitable.

Figure 3-11B shows a second type of DC excitation circuit. In this circuit the voltage regulator is a three-terminal IC voltage regulator ($U1$) of the LM-309, LM-340, 78xx or similar families. The selection of device depends upon the voltage required and the current normally drawn by the transducer. The load current is found from Ohm's law ($I = V/R$) where V is the regulator output voltage and R is the resistance of any one transducer element. In a typical case, the transducer will use a +5 volt excitation potential. If the resistance of the 'R' elements is >50 ohms, then by Ohm's law the current will be less than 100 mA. A 100 mA LM-309H, LM-340H, 78Lxx and so forth can be used.

The zener diode ($D1$) in Fig. 3-11B is not used for voltage regulation, but rather for overvoltage protection of the transducer. If the regulator ($U1$) fails, then +8 to +16 volts from the $V+$ line will be applied to the transducer. The purpose of $D1$ is to clamp the voltage to a value that is greater than the excitation voltage, but less than the maximum allowable excitation voltage rating of the transducer. A small fuse is sometimes inserted in series with the input (pin no. 1) of $U1$. The value of this fuse is set to roughly twice the current requirements (V/R) of the transducer, and will blow if the zener diode voltage is exceeded. The fuse adds a certain amount of protection. In other cases, a light emitting diode (LED) in series with diode $D1$ informs the

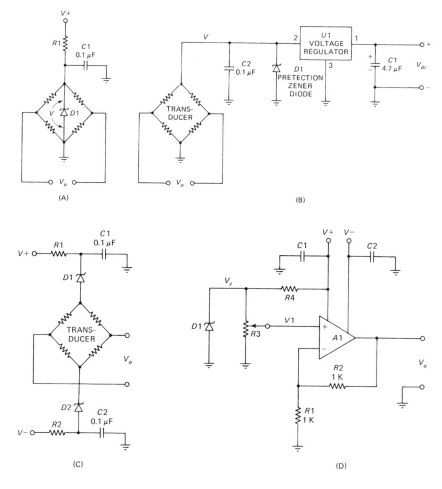

FIGURE 3-11 Bridge DC excitation circuits: (A) zener diode; (B) IC voltage regulator; (C) dual polarity zener diode; (D) op-amp voltage reference circuit.

user of a fault. Otherwise, the error would be interpreted simply as a higher transducer output signal which is indistinguishable from a higher value due to increased stimulus.

Some applications require a dual-polarity power supply. Figure 3-11C shows an excitation circuit in which two zener diodes are used, one each for positive and negative polarities. A certain amount of thermal stability is achieved by having opposite polarity power supplies because under temperature changes the voltage difference remains relatively constant.

Neither the zener circuit nor the three-terminal regulator circuit will deliver highly accurate output voltages. The voltage will be stable but not accurate. A typical three-terminal IC voltage regulator output voltage, for example, may vary several percent from sample to sample. If we need an accurate excitation voltage, then a circuit such as Fig. 3-11D might be used. This circuit is an operational amplifier voltage reference circuit in which the op-amp is a high current model, such as the LM-13080.

The output voltage from Fig. 3-11D is determined by R1, R2, the setting of R3 and the value of zener diode D1. The voltage V will be:

$$V = V1 \times \left[\frac{R2}{R1} + 1\right] \qquad (3\text{-}17)$$

or, since $R1 = R2 = 1$ kohm,

$$V = 2 \times V1 \qquad (3\text{-}18)$$

Voltage V1, at the noninverting input of A1, is a fraction of the zener voltage that depends upon the setting of potentiometer R3. We can adjust V1 from 0 Vdc to V_z Vdc, so the transducer voltage can be set at any value from 0 to $2V_z$ volts DC. In most cases, V will be set at 5.00 volts, 7.50 volts or 10.00 volts depending upon the design of the transducer.

Transducer amplifiers. The basic DC differential amplifier is the most commonly used circuit for amplifying Wheatstone bridge transducer signals. Such amplifiers are easily constructed from simple operational amplifiers (Fig. 3-12 shows such a circuit). Assuming that $R1 = R2$ and $R3 = R4$, the gain of the amplifier will be $R4/R2$, or $R3/R1$. The amplifier output voltage will be found from:

$$V_o = V_{in} \times \frac{R3}{R1} \qquad (3\text{-}19)$$

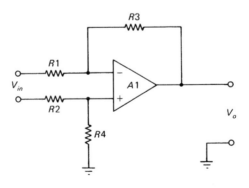

FIGURE 3-12 DC differential amplifier circuit.

84   MEASUREMENT AND INSTRUMENTATION CIRCUITS

where:
- V_o is the amplifier output voltage
- V_{in} is the transducer output voltage
- $R1$ and $R3$ are the resistors in the amplifier circuit

The gain required from the amplifier is determined from a *scale factor*, SF, which is the ratio between the voltage representing full scale at the output of the amplifier, and the voltage representing full scale at the output of the transducer:

$$SF = \frac{V_{o(max)}}{V_{in(max)}} \qquad (3\text{-}20)$$

where:
- $V_{o(max)}$ is the voltage V_o representing full scale
- $V_{in(max)}$ is the transducer output voltage V_{in} that represents full scale

From earlier you may recall that the transducer output voltage is found from the excitation voltage, the applied stimulus, and the sensitivity factor (ψ). The sensitivity factor is given in terms of millivolts (or microvolts) output per volt of excitation potential per unit of applied stimulus:

$$\psi = \frac{V_o}{V \times Q} \qquad (3\text{-}21)$$

The output voltage from the transducer is therefore found from:

$$V_{in} = P \times V \times \psi \qquad (3\text{-}22)$$

where:
- V_{in} is the transducer output voltage (or amplifier input voltage)
- V is the excitation voltage
- Q is the applied stimulus (force, pressure, etc.)

EXAMPLE 3-3

A chemical requires a 0 to 1000 Torr fluid pressure transducer and amplifier. The transducer is rated at a sensitivity factor (ψ) of 50 µV/V/Torr, and a range of −200 to +1200 Torr. The available excitation source was 6.95 volts. Calculate: (a) the full-scale output voltage, and (b) the gain required for an output voltage scale factor of 10 mV/Torr.

Solution

(a) At the required full scale pressure (1000 Torr) the output voltage is:

$$V_o = \psi V Q$$

$$V_o = \frac{50\ \mu V}{V \times T} \times (5.00\ \text{volts}) \times (1000\ \text{Torr})$$

$$V_o = 250\,000\ \mu V \times \frac{1\ mV}{1000\ \mu V} = 250\ mV$$

(b) The amplifier output voltage required will depend upon the desired display method. For example, a strip chart recorder might have a full-scale voltage range of 0.5 volts, 1.0 volt or some such value. Alternatively, a digital panel meter (DPM) may be used for the output display. Most DPMs have either a 0 to 1999 millivolt range or a 0 to 19.99 volt range, so a great deal of utility is gained by making the output voltage at full scale numerically the same at the DPM reading — for example 1000 Torr being represented by either 1000 millivolts or 10.00 volts. In that case, the DPM scale factor would be either 1 mV/Torr or 10 mV/Torr. The value of $V_{o(max)}$ in the equation above is 1000 mV. The gain of the amplifier is the scale factor SF described earlier:

$$SF = \frac{V_{o(max)}}{V_{in(max)}}$$

$$SF = \frac{10\,000 \text{ mV}}{250 \text{ mV}} = 40 \qquad \blacksquare$$

A practical design example. Physiologists and other life scientists sometimes use a strain gage force–displacement transducer such as the Grass model FT-3 shown in Fig. 3-13A for their experiments. Small displacements of the operating arm at one end of the transducer housing produces changes in the internal Wheatstone bridge that, in turn, result in an output voltage, V_{ot}. The sensitivity factor (ψ) of the transducer is 35 µV/V/gram-force [NOTE: the gram-force is a force equal to the force of gravity on a mass of one gram, i.e. 1 g-F = 980 dynes]. Forces to 10 g-F can be accommodated with excitation potential up to +7.5 Vdc.

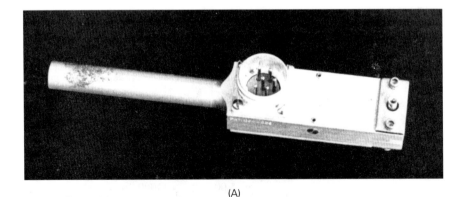

(A)

FIGURE 3-13 (A) Grass FT-3 strain gage transducer; (B) Wheatstone bridge amplifier circuit; (C) amplifier in box; (D) attached to transducer.

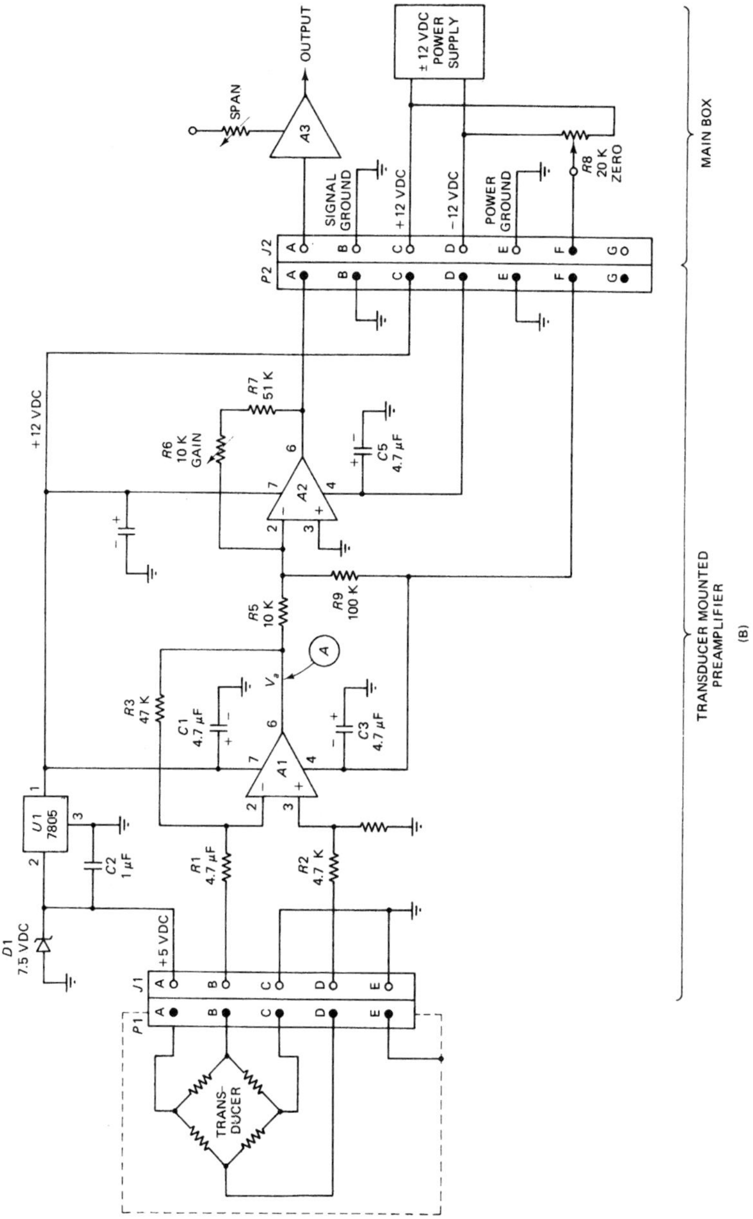

FIGURE 3-13 (continued)

MEASUREMENT AND INSTRUMENTATION CIRCUITS **87**

(C)

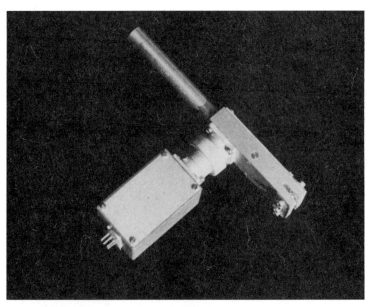

(D)

FIGURE 3-13 (*continued*)

A physiologist attempted to use an FT-3 transducer without a preamplifier because the manufacturer's offering was too expensive for the scientist's research budget. The experiment was to measure the minute contractions of a guinea pig heart *in vitro* in response to a certain stimulus. The transducer output, therefore, was a weak pulsatile voltage. The transducer was excited from a 6 Vdc battery. The balanced output of the transducer was used to drive the Channel A and Channel B inputs of an oscilloscope. When the 'scope is used in the A-B input mode, it will poorly mimick a differential amplifier. Unfortunately, gain differences between the two channels resulted in a common mode rejection ratio (CMRR) problem. In addition, the lead wires were unshielded, which led to excessive 50/60 Hz hum pick-up from surrounding AC power wiring in the laboratory. The problem was exacerbated by the fact that transducer signals were low compared with the 50/60 Hz induced signal.

It was decided to design and build a transducer preamplifier for the experiment. The excitation voltage was standardized to +5 Vdc because 5 volt IC voltage regulators were easily obtained. For a maximum 4 g-F stimulus, which the scientist believed to be the absolute maximum that would be seen, the transducer output voltage is:

$$V_{ot} = PVF \qquad (3\text{-}23)$$

$$V_{ot} = \frac{35\ \mu V}{V \times \text{g-F}} \times (5.00\ \text{Vdc}) \times (4\ \text{g-F})$$

$$V_{ot} = 35\ \mu V \times 5 \times 4 = 700\ \mu V$$

Converting to volts:

$$V_{ot} = 700\ \mu V \times \frac{1\ \text{volt}}{10^6\ \mu V} = 0.0007\ \text{volts}$$

Overcoming the 50/60 Hz interference problem required boosting the signal at the source to a higher level so that the induced 50/60 Hz signal is only a tiny fraction of the desired signal. In other words, the *signal-to-noise ratio* (S/N) must be improved. A strategy was adopted to boost V_{ot} to 10 mV/g-F, or 40 mV when the maximum 4 g-F is applied to the transducer. The required voltage gain is:

$$A_{vt} = \frac{V_{ota}}{V_{ot}} \qquad (3\text{-}24)$$

$$A_{vt} = \frac{\left[40\ \text{mV} \times \dfrac{1\ \text{volt}}{1000\ \text{mV}}\right]}{0.0007\ \text{volt}}$$

$$A_{vt} = \frac{0.04}{0.0007} = 57.14$$

A two-stage preamplifier circuit (Fig. 3-13B) was designed and built. It used a gain-of-ten ($A_{vt} = 10$) DC differential input stage ($A1$) to accommodate the balanced output of the Wheatstone bridge. The output signal of $A1$ (V_a) is boosted 10X to 0.007 volts (7 mV) when a 4 g-F is applied to the transducer.

Input resistors $R1$ and $R2$ must be at least ten times the Thevenin's looking-back resistance of the transducer, which for a Wheatstone bridge with all elements equal is the resistance of any one element. For the Grass FT-3 transducer used here the resistance elements were 200 ohms each, so any input resistor value >2000 ohms is acceptable. A value of 4.7 kohms was selected for reasons of convenience. The gain A_{v1} of this stage is found from:

$$A_{v1} = \frac{R3}{R1} \tag{3-25}$$

so,

$$R3 = A_{v1} R1 \tag{3-26}$$

Inserting the actual values:

$$R3 = (10)(4.7 \text{ kohms}) = 47 \text{ kohms}$$

A variable gain stage ($A2$) is used to boost V_{ota} to 40 mV when $V_a = 7$ mV. The gain required of $A2$ is:

$$A_{v2} = \frac{V_{ota(max)}}{V_{a(max)}} \tag{3-27}$$

or, with the values inserted:

$$A_{v2} = \frac{40 \text{ mV}}{7 \text{ mV}} = 5.714$$

Another way to calculate A_{v2} is to divide the total required gain by A_{v1}:

$$A_{v2} = \frac{A_{vt}}{A_{v1}} \tag{3-28}$$

$$A_{v2} = \frac{57.14}{10} = 5.714$$

A value of 10 kohms was selected for $R5$ for the terribly practical reason that it is a standard value, and thus is easy to obtain. The value had to be >1000 ohms in order to accommodate the normal <100 ohms output impedance of the operational amplifier. For a gain of 5.714 the feedback resistance $R_f = (R6 + R7)$ must be:

$$R6 + R7 = A_{v2} R5 \tag{3-29}$$

$$R6 + R7 = (5.714)(10 \text{ kohms}) = 57.14$$

Although a single precision resistor can be obtained to accommodate the required resistance (with an acceptable tolerance), the solution adopted here is to break up the total resistance between two standard values. For example, $R6$ is set to 10 kohms (and is a potentiometer), while $R7$ is a 51 kohm fixed resistor.

The circuit of Fig. 3-13B was built inside a small aluminum instrument box (see Fig. 3-13C) that was fitted with a mate to the transducer output connector. This configuration allowed the preamplifier to be mounted directly to the transducer housing (Fig. 3-13D).

A control box was connected to the preamplifier through P2/J2. This external box provides additional gain, as required. A gain-of-10 amplifier will boost the signal to a healthy 400 mV. The external box also provides an offset null (or 'ZERO') circuit, and gain control ('SPAN') to calibrate the system.

3-3.3 Microelectrodes

The 'microelectrode' is an ultra-fine electrical probe used to measure biopotentials at the cellular level. The microelectrode penetrates a cell that is immersed in an 'infinite' fluid (such as physiological saline) that, in electrical terminology, serves as the common or ground for the measurement. The fluid is, in turn, connected to a *reference electrode* (Fig. 3-14A). Although several types of microelectrode exist, the *metallic tip contact* is perhaps the most common because it is easily made in the research lab or an associated shop. The other form is a fluid-filled type in which the potassium chloride (KCL) column is used. In both types of microelectrode, an exposed contact surface of one to two micrometers (1 μm = 10^{-6} m) is in contact with the cell. As might be expected, this small surface area makes microelectrodes very high impedance devices, which require very high input impedance amplifiers.

Figure 3-14B shows the construction of a typical glass–metal microelectrode. A very fine platinum or tungsten wire is slip-fit through a 1.5 to 2 mm glass pipette. The tip of the pipette is acid etched, and then fire formed into the shallow angle taper shown. The microelectrode can then be connected to one input of the differential amplifier. There are two subcategories of this type of electrode. In one type, the metallic tip is flush with the end of the pipette taper, while in the other there is a thin layer of glass covering the metal point. This glass layer is so thin that thickness is usually measured in ångströms (the glass tip substantially increases the electrical impedance of the device).

Figure 3-14C shows a simplified equivalent circuit for the microelectrode (disregarding the contribution of the reference electrode). Analysis of this circuit reveals a problem that is due to the *RC* components. Resistor $R1$ and capacitor $C1$ are required because of the effects at the electrode–cell interface, and are frequency dependent. These values fall off to a negligible point at a rate of $1/(2\omega F)^2$, and are generally considerably lower than R_s and $C2$.

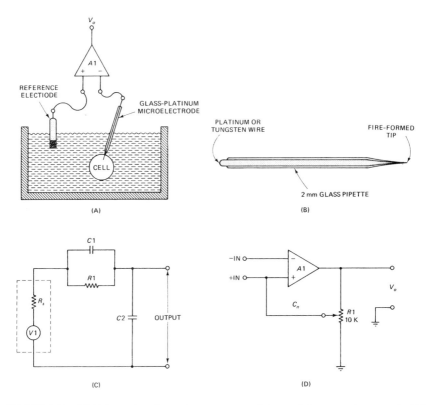

FIGURE 3-14 (A) Using microelectrode to measure living cell transmembrane potential; (B) pipette microelectrode; (C) equivalent circuit; (D) capacitance neutralization circuit.

Resistance R_s in Fig. 3-14C is the spreading resistance of the electrode, and is a function of the tip diameter. The value of R_s in metallic microelectrodes without the glass coating is approximated by:

$$R_s = \frac{\rho}{4 \times R} \qquad (3\text{-}30)$$

where:
 R_s is the resistance in ohms (Ω)
 ρ is the resistivity of the 'infinite' solution outside of the electrode
  (e.g. 70 ohm cm for physiological saline)
 r is the tip radius (typically 0.5 µm for a 1 µm electrode)

EXAMPLE 3-4

Assuming the typical values given above, calculate the tip spreading resistance of a 1 µm microelectrode.

Solution

$$R_s = \frac{\rho}{4\pi r}$$

$$R_s = \frac{70 \ \Omega \ \text{cm}}{4\pi 0.5 \ \mu\text{m} \times \frac{10^{-4} \text{cm}}{1 \ \mu\text{m}}} = 111\,408 \text{ ohms}$$

The impedance of glass-coated metallic microelectrodes is at least one or two orders of magnitude higher than this figure. ∎

For fluid-filled KCL microelectrodes with small taper angles ($\pi/180$ radians), the series resistance is approximated by:

$$R_s = \frac{2\rho}{\pi r a} \tag{3-31}$$

where:
 R_s is the resistance in ohms (Ω)
 ρ is the resistivity (typically 3.7 Ω cm for 3M KCL)
 r is the tip radius (typically 0.1 μm)
 a is the taper angle (typically $\pi/180$)

EXAMPLE 3-5
Find the series impedance of a KCL microelectrode using the values shown above:

Solution

$$R_s = \frac{2\rho}{\pi r a}$$

$$R_s = \frac{(2)(3.7 \ \Omega \ \text{cm})}{(3.14) \times (0.1 \ \mu\text{m}) \times \frac{10^{-4} \text{ cm}}{1 \ \mu\text{m}} \frac{3.14}{180}} = 13.5 \text{ megohms} \quad ∎$$

The capacitance of the microelectrode is given by:

$$C2 = \frac{0.55 e}{\ln(R/r)} \frac{\text{pF}}{\text{cm}} \tag{3-32}$$

where:
 e is the dielectric constant of glass (typically 4)
 R is the outside tip radius
 r is the inside tip radius (r and R in same units)

EXAMPLE 3-6
Find the capacitance ($C2$ in Fig. 3-14C) if the pipette radius is 0.2 μm and the inside tip radius is 0.15 μm.

Solution

$$C2 = \frac{0.55e}{\ln(R/r)} \frac{\text{pF}}{\text{cm}}$$

$$C2 = \frac{(0.55)(4)}{\left[\ln \dfrac{0.2 \ \mu\text{m}}{0.15 \ \mu\text{m}}\right]} \frac{\text{pF}}{\text{cm}} = 7.7 \ \frac{\text{pF}}{\text{cm}} \qquad \blacksquare$$

How do these values affect performance of the microelectrode? Resistance R_s and capacitor $C2$ operate together as an RC low-pass filter. For example, a KCL microelectrode immersed in 3 cm of physiological saline has a capacitance of approximately 23 pF. Suppose it is connected to the amplifier input (15 pF) through 3 feet of small diameter coaxial cable (27 pF/ft, or 81 pF). The total capacitance is $(23 + 15 + 81)$ pF $= 119$ pF. Given a 13.5 megohm resistance, the frequency response (at the -3 dB point) is:

$$F_{\text{Hz}} = \frac{1}{2\pi RC} \qquad (3\text{-}33)$$

where:
F_{Hz} is the -3 dB point in hertz (Hz)
R is the resistance in ohms (Ω)
C is the capacitance in farads (F)

EXAMPLE 3-7

For $C = 119$ pF (1.19×10^{-10} farads) and $R = 1.35 \times 10^7$ ohms, find the frequency response upper -3 dB point.

Solution

$$F_{\text{Hz}} = \frac{1}{2\pi RC}$$

$$F_{\text{Hz}} = \frac{1}{(2)(3.14)(1.35 \times 10^7 \ \Omega)(1.19 \times 10^{-10} \ \text{F})}$$

$$F_{\text{Hz}} = 99 \ \text{Hz} \approx 100 \ \text{Hz} \qquad \blacksquare$$

A 100 Hz frequency response, with a -6 dB/octave characteristic above 100 Hz, results in severe rounding of the fast risetime *action potentials* found in biological cells. A strategy must be devised in the instrument design to overcome the effects of capacitance in high impedance electrodes.

Neutralizing microelectrode capacitance. Figure 3-14D shows the standard method for neutralizing the capacitance of the microelectrode and associated circuitry. A neutralization capacitance, C_n, is in the positive feedback path along with a potentiometer voltage divider. The value of this capacitance is:

$$C_n = \frac{C}{A_v - 1} \qquad (3\text{-}34)$$

where:
C_n is the neutralization capacitance
C is the total input capacitance
A_v is the gain of the amplifier

EXAMPLE 3-8

A microelectrode and its cabling exhibit a total capacitance of 100 pF. Find the value of neutralization capacitance (Fig. 3-14D) required for a gain-of-10 amplifier.

Solution

$$C_n = C/(A - 1)$$
$$C_n = (100 \text{ pF})/(10 - 1)$$
$$C_n = 100 \text{ pF}/9 = 11 \text{ pF} \qquad \blacksquare$$

3-3.4 Light transducers

Various classes of instruments use light as a transducible element. Colorimetry instruments are often based on transducers using this principle. Various instrumentation techniques may depend on various properties of light and light transducers. In some cases, only the existence or nonexistence of the light beam is important. Examples of this use is the PAPER OUT sensor on a computer printer, or the light beam that counts entrances and exits from a building or controlled space. In other cases, the color of the light is important. In still others, the absorption of particular colors is the important factor. In all of these cases the light beam must be sensed before it can be used.

Light is a form of electromagnetic radiation, and in the ultimate sense is the same as radio waves, infrared (heat) waves, ultraviolet, and X-rays. The principal differences between these types of electromagnetic radiation are the frequency and wavelength. The wavelength of visible light is 400 to 800 nanometers (1 nm = 10^{-9} meters); infrared has longer wavelengths than visible light and ultraviolet has wavelengths shorter than visible light; X-radiation has wavelengths even shorter than ultraviolet. Frequency and wavelength are related in electromagnetic radiation by the equation:

$$\lambda = \frac{c}{f} \qquad (3\text{-}35)$$

where:
c is the velocity of light (300 000 000 meters/second)
λ is wavelength in meters (m)
f is frequency in hertz (Hz)

From the above equation you can see that light has a frequency on the order of 10^{14} Hz (compare with the frequencies of AM and FM broadcast bands in the radio portion of the spectrum, i.e. 10^6 Hz and 10^8 Hz, respectively).

Because IR, UV and X-radiation are similar in nature and close in wavelength to visible light, many of the sensors and techniques applied to visible light also work to some extent in those adjacent regions of the electromagnetic spectrum. Although performance varies somewhat, and some photosensitive devices aren't even useful in those areas of the spectrum, it is nonetheless true that designers dealing with those spectra may find some of these devices useful.

The photosensors described in this section depend upon quantum effects for their operation. Quantum mechanics arose as a new idea in physics in December 1900, the very dawn of the 20th century, with a now-famous paper by German physicist Max Planck. He had been working on thermodynamics problems resulting from the fact that the experimental results reported in 19th-century physics laboratories could not be explained by classical Newtonian mechanics. The solution to the problem turned out to be a simple, but terribly revolutionary idea: energy existed in discrete bundles, not as a continuum. In other words, energy comes in packets of specific energy levels; other energy levels are excluded. The name eventually given to these energy packets was *quanta*, thus was born the branch of physics called *quantum mechanics*. The name given to energy bundles that operated in the visible light range was *photons*.

The energy level of each photon is expressed by the equation:

$$E = \frac{Ch}{\lambda} \tag{3-36}$$

or alternatively,

$$E = h\nu \tag{3-37}$$

where:
  E is the energy in electron volts (eV)
  c is the velocity of light in meters per second
  λ is the wavelength in meters (m)
  h is Planck's constant (6.62×10^{-34} J s)
  ν is the frequency of light waves in hertz (Hz)

(Note: the constants ch are sometimes combined, and expressed together as 1240 electron volts per nanometer (eV/nm).)

The basis of some light sensors is to construct a device that allows at least one electron to be freed from its associated atom by one photon of light. Materials in which the electrons are too tightly bound for light photons to do this work will not work well as light sensors.

The phototube. The *phototube* is a vacuum tube device that is based on the *photoelectric effect*. Interestingly, physicist Albert Einstein won the Nobel Prize in physics for his explanation of the photoelectric effect, not for either Special or General Theories of Relativity as is commonly assumed. Einstein wrote three seminal papers for the 1905 edition of *Annalen der Physik*, any of which could have qualified him for the Nobel prize: explanation of Brownian motion as a molecular effect, explanation of the photoelectric effect, and Special Relativity.

The effect that Einstein explained in 1905 had long perplexed physicists. If you shine a light onto certain types of metallic plate in a vacuum electrons are emitted from the surface of the plate. Oddly, increasing the brightness of the light does not increase the energy level of the emitted electrons. If this phenomena were purely a mechanical kinetic event, then one would assume from the classical point of view that increasing the intensity of the light would increase the energy of the electrons emitted from the surface. It turned out, however, that changing the *color* of the light affected the energy level of the emitted electrons. Red light produced lower energy electrons than violet light. The *energy level of the electrons is color sensitive*, even though the *amount of current (i.e. number of electrons) emitted is intensity sensitive*. Once Planck's principle was known, however, Einstein was able to explain this effect by quantum principles. From the above equations you can see that the higher frequency (shorter wavelength) violet colored light is significantly more energetic than red light.

Figure 3-15A shows a phototube based on the photoelectric effect. The cathode is a metallic plate that is exposed to light; it is made of a material that will easily emit electrons when light is applied. The anode is positively charged (with respect to the cathode) by an external DC power supply. Electrons emitted from the photocathode are collected by this anode, and can be read on an external meter (or other circuit).

The photoemission process is less efficient than is needed in many cases. We can make the system more efficient by using a *photomultiplier tube* (Fig. 3-15B). In this type of photosensor there are a number of positively charged anodes, called *dynodes*, that intercept the electrons. Electrons are emitted when light impinges the cathode. They are accelerated through a high voltage potential ($V1$) to the first dynode. They acquire substantial kinetic energy during this transition, so when each electron strikes the metal it gives up its kinetic energy. Because of the conservation of energy principal, giving up kinetic energy requires some energy to be converted to heat, while some is converted to additional kinetic energy by dislodging other electrons from the dynode surface. Thus, a single electron causes two or more additional electrons to be dislodged. These electrons are accelerated by another high voltage potential ($V2$), and produces the same effect at the second dynode. The process is repeated several times, and each time several more electrons

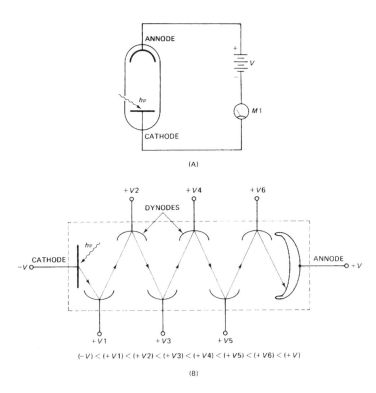

FIGURE 3-15 (A) photoelectric tube; (B) photomultiplier tube.

join the process for each previously accelerated electron. Finally, the electron stream is collected by the anode, and can be used in an external circuit.

Photovoltaic cells. A *photovoltaic cell* is one in which a potential difference is generated, and thus a current flow, by shining light onto its surface. The common solar cell is an example of the photovoltaic cell. There are only a few instrumentation applications for this type of cell. Figure 3-16 shows a typical circuit for instrumentation applications of the photovoltaic cell. The cell is connected across the input of a high impedance amplifier, such as the noninverting operational amplifier shown. The output voltage is found from:

$$V_o = V1 \times \left[\frac{R2}{R1} + 1 \right] \qquad (3\text{-}38)$$

Photoresistors (photoconductive cells). A *photoresistor* is a device that changes electrical ohmic resistance when light is applied. Figure 3-17 shows the usual circuit symbol for photoresistors; it is the normal resistor symbol enclosed within a circle, and given the greek letter *lambda* symbol to denote

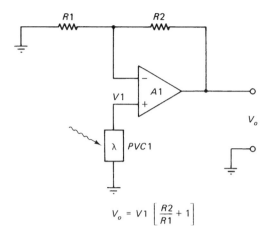

FIGURE 3-16 Photovoltaic cell amplifier.

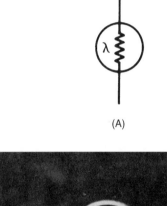

(A)

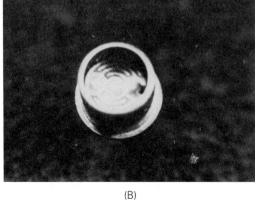

(B)

FIGURE 3-17 (A) Photoconductive cell (photoresistor) schematic symbol; (B) photoresistor cell.

that it is a resistor that responds to light. When photoresistors are specified, both a *dark resistance* and a *light/dark resistance ratio* are listed. In most common varieties, the resistance is very high when dark, and drops very low under intense light. The intensity of the light affects the resistance, so these devices can be used in photographic lightmeters, densitometers, colorimeters and so forth.

Figure 3-18 shows three circuits in which photoresistors can be used. The half-bridge circuit is shown in Fig. 3-18A. In this circuit, the photoresistor is connected across the output of a voltage divider made up of $R1$ and $PC1$. The output voltage is given by:

$$V_o = \frac{V\ PC1}{R1 + PC1} \tag{3-39}$$

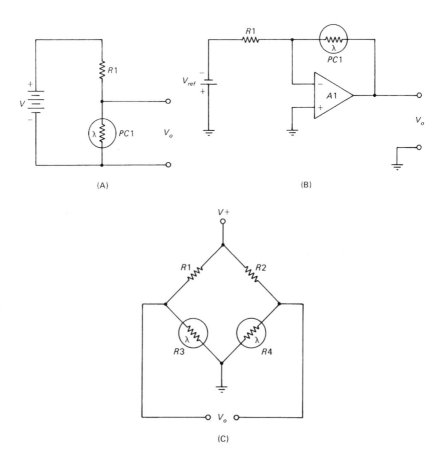

FIGURE 3-18 Photoconductive cell circuits: (A) half-bridge; (B) op-amp inverting follower; (V) Wheatstone bridge.

where:

    V_o is the output potential
    V is the applied excitation potential
    $R1$ and $PC1$ are in ohms

A problem with this circuit is that the output potential does not ever drop to zero, but always has an offset value.

A second way to use the photoresistor is shown in Fig. 3-18B. Here the photoresistor is the feedback resistor in an operational amplifier inverting follower circuit. The output voltage, V_o, is found from:

$$V_o = -(-V_{ref}) \times \frac{PC1}{R1} \qquad (3\text{-}40)$$

The circuit of Fig. 3-18B provides a low-impedance output, but like the half-bridge circuit (Fig. 3-18A), the output voltages does not drop to zero. In addition, with some photoresistors the dynamic range of the operational amplifier may not match the dark/light ratio of the photoresistor at practical values of $-V_{ref}$ potential.

The last photoresistor configuration is the Wheatstone bridge shown in Fig. 3-18C. This circuit allows the output voltage to be zero under the right circumstances, and is the circuit favored by most designers. If a low impedance output is required, or additional amplification is needed, then a DC differential amplifier can be connected across output potential.

Photodiodes and phototransistors. Perhaps the most modern light sensor is the PN junction configured in the form of special photodiodes or phototransistors. In certain types of diode, the level of the reverse leakage current increases when the junction is illuminated. Figure 3-19A shows the basic circuit used for these sensors. The diode is normally reverse biased, with a current-limiting resistance in series. Microammeter $M1$ measures the reverse leakage current that crosses the PN junction during this type of operation. When light strikes the PN junction, the reading on $M1$ will increase.

The same principle applies to a class of NPN or PNP transistors called *phototransistors* (Fig. 3-19B). In these devices, collector to emitter current flows when the base region is illuminated. These devices are the heart of optoisolator and optocoupler integrated circuits, as well as being used in various instrumentation sensor applications.

Figures 3-19C and 3-19D show one way in which photodiodes and phototransistors are used. The current flow through the device under light conditions is the transducible event, so they must be connected into a circuit that makes use of that property. The inverting follower operational amplifier circuit works well. The output voltage V_o is equal to the product of the input current and the feedback resistance ($I_{in} \times R_f$). In the case of Fig. 3-19D, a *zero control* is added, and can also be added to the other circuit as well.

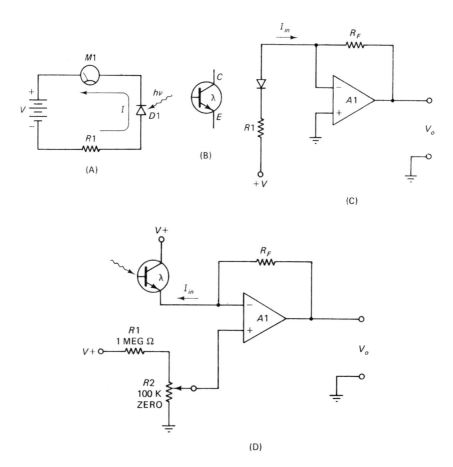

FIGURE 3-19 (A) Photodiode circuit; (B) photodiode schematic symbol; (C) op-amp photodiode amplifier; (D) photodiode amplifier with offset compensation.

Photocolorimetry. One of the most basic forms of instrument is also both the oldest and most commonly used: *photocolorimetry*. These devices are used to measure oxygen content in blood, CO_2 content of air, water vapor content, blood electrolyte (Na and K) levels, and a host of other measurements in the medical laboratory.

Figure 3-20 shows the basic circuit of the most elementary form of colorimeter. Although the circuit is very basic, this is the very circuit of a widely used medical blood oxygen meter. The circuit is a Wheatstone bridge that uses a pair of photoresistors as the transducing elements. Potentiometer R5 is used as a balance control, and it is adjusted for zero output ($V_o = 0$) when the light shines equally on both photoresistors. Recall from the Chapter 1 discussion, that output voltage V_o will be zero when the two legs of the bridge are balanced. In other words, V_o is zero when $R1/R2 = R3/R4$. It

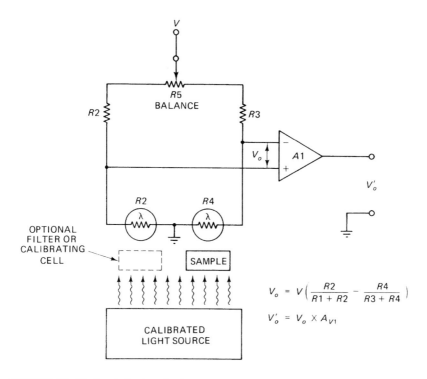

FIGURE 3-20 Photocolorimeter circuit.

is not necessary for the resistor elements to be equal (although that is usually the case), only that their ratios be equal. Thus, a 500k/50k ratio for $R1/R2$ will produce zero output voltage when $R3/R4 = 100k/10k$.

The photoresistors are arranged so that light from a calibrated source shines on both equally and fully, except when an intervening filter or sample is present. Thus, the bridge can be nulled to zero using potentiometer $R5$ under this zero condition. In most instruments, a translucent sample is placed between the light source and one of the photocells. The comparison of light transmission through air (or a vacuum or a specified wavelength filter) to light transmitted through the sample being tested is a measure of sample density, and is thus transducible. Consider some practical applications below.

Blood O_2 Level. An old (but still used) method for measuring blood oxygen level is the colorimeter of Fig. 3-20. It works on the fact that the redness of blood is a measure of its oxygenation. This instrument is nulled with neither standard filter cell nor blood in the light path. A standard 800 nm filter cell is then introduced between the light source and $R2$, and a blood sample is placed in a standardized tube between the light source and $R4$. The degree of blood saturation in the sample is thus reflected by the difference in the bridge reading. On one model, a separate resistor across the $R1/R2$

arm is used to bring the bridge back into null condition, and the dial for that resistor is calibrated in percent O_2. More modern instruments based on digital computer techniques provide the measurement in a more automatic manner.

Respiratory CO_2 level. The exhaled air from humans is roughly 2 to 5% carbon dioxide, while the percentage of CO_2 in room air is negligible. A popular form of end tidal CO_2 meter is based on the fact that CO_2 absorbs infrared (IR) waves. The 'light source' in that type of instrument is actually either an IR LED or a *Cal-Rod*® (identical to the type used to heat electrical coffee pots). The photocells are selected for IR response. In this type of instrument, room air is passed through a *cuvette* placed between $R2$ and the heat source, while patient expiratory air is passed through the same type of *cuvette* that is placed between the heat source and $R4$. The difference in IR transmission is a function of the percentage of CO_2 in the sample circuit.

The associated electronics will allow zero and maximum span (i.e. gain) adjustment. The zero point is adjusted with room air in both cuvettes, while the maximum scale (usually 5% CO_2) is adjusted with the sample cuvette purged of room air and replaced with a calibration gas (usually 5% CO_2 and 95% nitrogen).

Blood electrolytes. Blood chemistry includes levels of sodium (Na) and potassium (K). An instrument commonly used for these forms of measurement is the *flame photometer* (Fig. 3-21). This form of colorimeter replaces the light source with a flame produced in a gas carburetor. The sample is injected into the carburetor, and burned along with the gas/air mixture. The colors emitted on burning are proportional to the concentrations of Na and K ions in the sample. A special gas is used to burn cleanly with a blue flame when no sample is present. In medical applications, a specified size sample of patient's blood is mixed with a premeasured indium calibrating solution. The solution is well mixed, and then applied to the carburetor. By comparing the intensities of the colors generated by burning Na and K ions with the intensity of the calibration color, the instrument can infer the concentration of the two elements.

Photospectrometer. This class of instruments depends upon the fact that certain chemical compositions absorb wavelengths of light to different degrees. If you are familiar with the spectrum of sunlight created when the light passes through a prism, or is reflected from a diffraction grating, then you are familiar with the physical basis for the spectrophotometer (see Fig. 3-22). Either the light beam or the photodetector must move to examine each of the light wavelengths in turn. In most cases, the sample and the photodetector are kept stable, while a diffraction grating or prism is rotated to permit all colors of the spectrum to fall on the same photo detector. By comparing the amplitudes of the light at different angles, we can infer the transmission of the respective colors, and create a chart that shows amplitude versus wavelength.

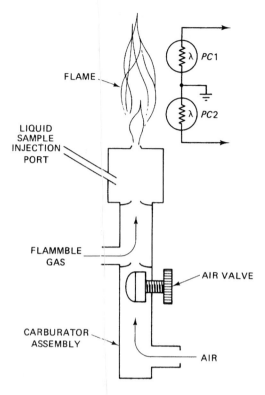

FIGURE 3-21 Flame photometer.

Ratiometric measurements. A significant problem with photometric measurements is keeping the light source constant in both intensity and color. The simple fact is that all forms of light emitter show change, and that change introduces artifact in the data — if it occurs between calibration and measurement. The answer to many of these problems is the use of *ratiometric* measurement technique (Fig. 3-23).

In a ratiometric system, the collimated light beam is passed through a 50% beam splitting prism. This optical device splits the beam into two equal amplitude beams at right angles to each other. The incident beam is passed through an optical path of length L to photosensor $PC2$; the sample beam also passes through an optical path of length L, but passes through the sample being measured as well. If the optical properties of the two paths are the same, and they are of the same length, then the light intensities arriving at the two colorimeter photosensors will be the same, except for any energy lost in the sample. Thus, the Wheatstone concept can be used. But first, some signal processing of the signal is needed. The circuit must take the *ratio* of $V1$ and $V2$, the outputs of $PC1$ and $PC2$, respectively, either in the computer, or in an analog multiplier. Thus, a change in the light level affects both photosensors

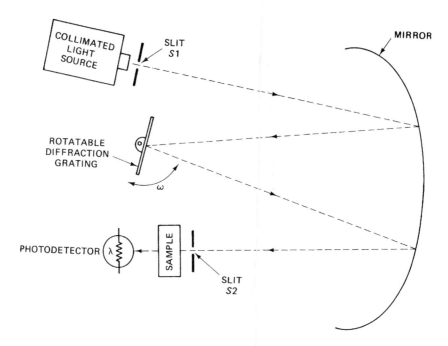

FIGURE 3-22 Photospectrometer.

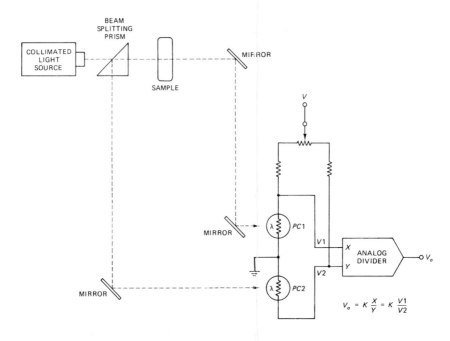

FIGURE 3-23 Ratiometric colorimeter circuit.

equally, so the only difference between the two is the properties of the sample. Analog logarithmic amplifiers can be used for this purpose.

A practical design example: the pH-meter. Chemists use pH as a measure of the relative acidity or alkalinity (base) of a fluid. If a solution is neither acid nor base then it is said to be neutral, and has a pH of 7.00. Acidic solutions have a pH < 7, while basic solutions have a pH > 7. Human blood has a normal pH of 7.4 ± 0.04 (i.e. 7.36 to 7.44), and pH values outside this range are considered pathological. A pH electrode is a special high impedance chemical glass electrode that produces an output voltage that is a function of fluid pH (Fig. 3-24A). A typical pH electrode will output 0 volts at pH = 7, plus or minus the offsets represented by a tolerance band and the electrode temperature.

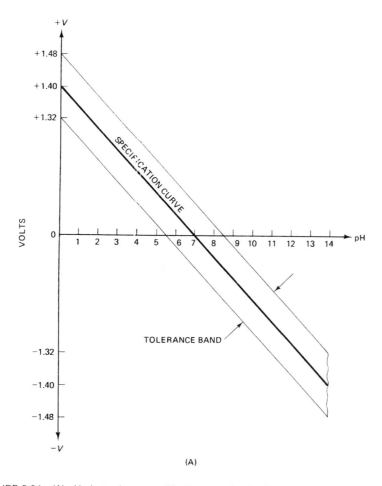

FIGURE 3-24 (A) pH electrode curve; (B) pH meter circuit; (C) calibration curve.

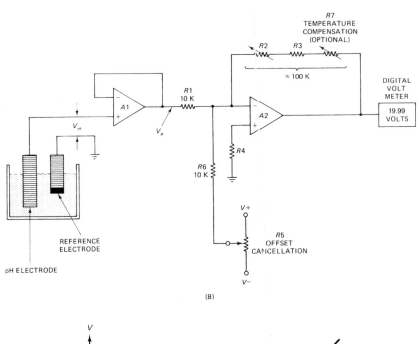

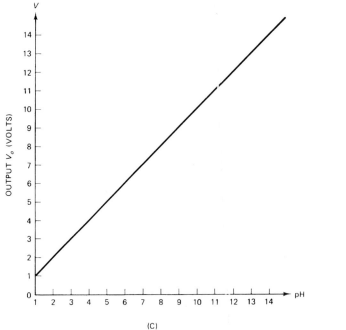

FIGURE 3-24 (*continued*)

108 MEASUREMENT AND INSTRUMENTATION CIRCUITS

In order to design a simple digital pH meter it is necessary to create a circuit that produces 1 Vdc output at pH = 1, 7 Vdc when pH = 7 and 14 Vdc when pH = 14.

The input impedance of the circuit must be >100 megohms because of the glass electrode source impedance. This requirement can be met by using a BiFET or BiMOS operational amplifier, or an electrometer amplifier. The unity gain configuration is used in the input stage shown in Fig. 3-24B.

The slope of the characteristic curve in Fig. 3-24A is negative, while the slope required by the design is positive (Fig. 3-24C). Therefore, an inverting stage (A2) is used to follow the input stage (A1). Accommodating electrode tolerances requires both an *offset control* and a variable voltage gain.

The offset control is adjusted to set V_o to +7.00 volts when the input voltage is zero ($V_{in} = 0$). The amplifier gain (A_{v2}) is set such that $V_o = +14.00$ volts when $V_{in} = -1.4$ volts, or

$$A_{v2} = \frac{V_{o(max)}}{V_{in(max)}}\bigg|_{pH=14} \qquad (3\text{-}41)$$

$$A_{v2} = \frac{14 \text{ volts}}{-1.4 \text{ volts}} = -10$$

Because the tolerance band is ±80 mV, the gain has to have a range of operation. The maximum gain:

$$A_{v2(max)} = \frac{14 \text{ volts}}{-1.32 \text{ volts}} = -10.61$$

... and the minimum gain:

$$A_{v2(max)} = \frac{14 \text{ volts}}{-1.48 \text{ volts}} = -9.45$$

The amplifier must have a gain that can be varied between -9.45 and -10.61. If $R1 = 10$ kohms, then $(R2 + R3)$ must be variable between:

$$(R2 + R3) = (10 \text{ kohms})(9.45) = 94.5 \text{ kohms}$$

and

$$(R2 + R3) = (10 \text{ kohms})(10.61) = 106.1 \text{ kohms}$$

A combination of a 91 kohms fixed resistor in series with a 20 kohms multi-turn potentiometer will solve the problem.

The offset control must set voltage V_b to 1.00 when $V_{in} = +1.4$ volts. Assuming a nominal gain of -10 for stage A2, and $+1$ for A1, voltage V_b would normally be $[(V_{in})(A_v)] = [(+1.4 \text{ Vdc})(-10)] = -14.00 \text{ Vdc}$. The potential at the wiper of the potentiometer (V_c) normally sees a nominal gain of -10 (i.e. $(R2 + R3)/R6$). The goal is to offset -14 Vdc to $+1$ Vdc, a total of +15 Vdc change. Given the gain of −10:

$$V_c = \frac{V_{b(desired)}}{A_v} \quad (3\text{-}42)$$

$$V_c = \frac{+15 \text{ Vdc}}{-10} = -1.5 \text{ Vdc}$$

Under the condition where the value of V_{in} is +1.4 volts (representing pH = 1), and $V_c = +1.3$ Vdc, then the total output voltage will be:

$$V_b = -(R2 + R3)\left[\frac{V_a}{R1} + \frac{V_c}{R6}\right] \quad (3\text{-}43)$$

$$V_b = -(100 \text{ kohms})\left[\frac{+1.4 \text{ volts}}{10 \text{ k}\Omega} + \frac{-1.5}{10 \text{ k}\Omega}\right]$$

$$V_b = -1$$

If a temperature compensation control is needed, then another offset control is added to the circuit at *A2*. This control is operable from the instrument front panel. In some instruments, the temperature offset potential is provided by measuring the temperature of the solution being measured. In at least one case, a thermistor or PN junction temperature sensor is embedded in the pH electrode housing. The student is left to think about the design of an automatic temperature compensation circuit when a 10 mV/K sensor is used, and the temperature range will be +20°C to +45°C (NOTE: 0°C = 273K). (Don't you just hate it when textbook authors do that?)

3-4 SUMMARY

1. Transducers are devices that convert one form of energy (e.g. pressure, displacement, force, etc.) to electrical energy for purposes of measurement, tabulation or control.
2. The phenomenon of *piezoresistivity* is often used for transducer. This phenomenon causes a change in resistance due to mechanical deformation of an electrical conductor.
3. A *strain gage* is a piezoresistive element that produces a resistance change in response to an applied force. Unbonded and bonded types are used.
4. Most strain gages and other resistive transducers are used in Wheatstone bridge circuits, or modified bridges.
5. A DC differential amplifier can be used to process the signal output from a Wheatstone bridge strain gage.
6. Light transducers can be photovoltaic cells, photoresistors, photodiodes or phototransistors. The latter two devices operate from the fact that PN junction leakage current is altered by light falling on the junction.

3-5 RECAPITULATION

Now return to the objectives and Pre-quiz questions at the beginning of the chapter and see how well you can answer them. If you cannot answer certain questions, place a check mark to each and review the appropriate parts of the text. Next, try to answer the questions and work the problems below, using the same procedure.

3-6 STUDENT EXERCISES

1. Design and build a transducer excitation source such as Fig. 3-11D to produce an output potential of 10.00 Vdc from an available zener diode (select V_z between 4.2 and 6.8 Vdc).
2. Using the exercise above as a guide, produce a circuit that offers *bipolar* ±10 Vdc outputs.
3. Design and build a gain-of-20 preamplifier for a microelectrode. Make provision for cancelling input capacitance (use a 20 foot length of coaxial cable between the signal generator and the amplifier to simulate electrode capacitance).
4. Design and build a photoelectric amplifier such as Fig. 3-18B. Select a photoresistor and a value of V_{ref} that will produce an output signal in response to light.
5. Build a Wheatstone bridge such as Fig. 3-18C using available photoresistors. Scale R1 and R2 to the nominal photoresistor resistance in room light. Connect the outputs to a DC differential amplifier with a gain of 10. Examine what happens to the output signal of the bridge when first one and then the other photoresistor is covered.
6. Design and build a pH meter using the circuit of Fig. 3-25 as a simulated pH electrode. Measure circuit performance using various settings of the SET pH control.
7. Design and build the amplifier to process the signals from the Wheatstone bridge transducer simulator shown in Fig. 3-26. The maximum output voltage when

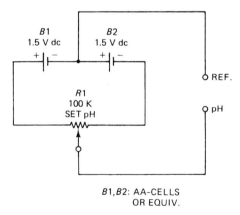

FIGURE 3-25 pH meter calibration circuit.

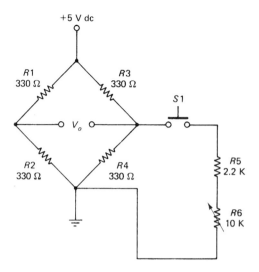

FIGURE 3-26 Instrument calibration circuit.

$R6 = 0$ ohms should be 2 volts. Measure the amplifier output (a) with $S1$ open, and (b) with $S1$ closed (at various settings of $R6$).

3-7 QUESTIONS AND PROBLEMS

1. Resistance change under mechanical deformation is called _____ .
2. A strain gage is constructed of a resistance wire held taut between two flexible supports. This is a _____ strain gage.
3. A strain gage is constructed of semiconductor resistance material cemented to a thin metal diaphragm. This is a _____ strain gage.
4. Draw a Wheatstone bridge circuit. If all elements are 300 ohms the output voltage will be _____ volts.
5. Calculate the Thevenin's equivalent 'looking back' resistance of a Wheatstone bridge transducer in which all four elements have a nominal resting value of 200 ohms.
6. A Wheatstone bridge consists one active strain gage element, with the other three resistances being fixed and equal to each other; the nominal resistance value of the strain gage (R) is equal to the value of the three fixed resistors. If an excitation potential of $+10.00$ Vdc is applied, what will the output voltage be if the strain gage element changes to a value of $0.9R$?
7. Work the above problem for the cases of (a) two active elements and (b) four active elements.
8. A force transducer is used to measure the weight of small objects. The transducer elements have a resting resistance of 250 ohms, and an excitation potential of $+6$ Vdc is applied. When an 800 mg mass is placed on the scale all four bridge arms change by 6.7 ohms. Calculate the output potential.

9. A Wheatstone bridge strain gage pressure transducer has a sensitivity factor of 48 μV/V/mmHg, and an optimum excitation potential of 6 Vdc. Calculate the output voltage from this transducer at the optimum excitation when a pressure of 120 mmHg is applied.

10. Using the parameters in the question above, determine the gain required in the transducer amplifier to output 1.20 Vdc when the static pressure applied to the transducer is 120 mmHg.

11. A transducer excitation source such as Fig. 3-11D is built using a 1.36 Vdc band-gap diode as the reference source. Calculate the output voltage when $R1 = 1000$ ohms and $R2 = 2700$ ohms.

12. The full-scale output potential of a transducer is 400 μV when a force of 2600 dynes is applied. Calculate the amplifier scale factor required to produce an output voltage of 2.6 Vdc when this force is applied.

13. Calculate the source resistance of a 0.8 μm diameter glass microelectrode when immersed in normal physiological saline solution ($\rho = 71$ ohm cm).

14. A microelectrode has a source resistance of 126 kohms, and a capacitance of 49 pF. Calculate the -3 dB frequency response.

15. Calculate the neutralization capacitance required to accommodate a 130 pF electrode in a gain-of-10 amplifier.

16. The phototube operates on the basis of the _____ effect.

17. A phototube that contains a number of special anodes called *dynodes* is a _____ tube.

18. A transducer produces an output voltage when light shines on it. This is a _____ cell.

19. Two photoresistors are connected into a Wheatstone bridge circuit. Light from a calibrated source falls on both photoresistors equally to calibrate, and then a sample is placed in the path between the light source and one photoresistor to make a measurement. This class of instrument is called a _____ .

20. A Wheatstone bridge such as Fig. 3-3 is constructed with the following components: $R1 = 10$ kohms, $R2 = 6.8$ kohms, $R3 = 6.8$ kohms, and $R4 = 5.6$ kohms. Calculate the output voltage V_o when a 1.5 Vdc battery is connected for V. What will the output voltage be if $R4$ is increased to 10 kohms.

21. Describe different methods of linearizing the output signal produced by a transducer.

22. Draw the circuit for a microelectrode amplifier that provides for neutralization of electrode and cable capacitances.

23. Draw the circuit for a Wheatstone bridge transducer in which both *balance* and *zero* controls are provided.

24. What is the purpose of zener diode $D1$ in Fig. 3-11B?

25. What are the most desirable properties of the input amplifier ($A1$) in Fig. 3-25B?

CHAPTER 4

Integrated circuit timers

OBJECTIVES
1. Understand the internal operation of the 555 timer and related IC timers.
2. Be able to design monostable multivibrator circuits based on the 555 timer.
3. Be able to design astable multivibrator circuits based on the 555 timer.
4. Understand the wide range of potential applications of the 555 timer and related devices.

4-1 PRE-QUIZ

These questions test your prior knowledge of the material in this chapter. Try answering them before you read the chapter. Look for the answers (especially those you answered incorrectly) as you read the text. After you have finished studying the chapter try answering these questions again, and those at the end of the chapter (see Section 4-12).

1. Calculate the duration of a 555 monostable multivibrator in which $R1 = 56$ kohms and $C1 = 0.001$ µF.
2. List the pinouts of the 555 timer and briefly describe the operation of each.
3. An astable multivibrator made from a 555 IC timer uses the following components: $R1 = 68$ kohms, $R2 = 33$ kohms, and $C1 = 0.0$ µF. Calculate: (a) the duty cycle, and (b) the operating frequency.
4. A 555 timer is operated from a $+12$ Vdc regulated DC power supply. What is the minimum value of the load resistor that can be connected between pin no. 3 on the 555 and $V+$?

4-2 INTRODUCTION TO THE 555-FAMILY IC TIMERS

The integrated circuit timer represents a class of chips that are extraordinarily well behaved and easy to apply. These timers are based on the properties of the series *RC* timing network and the op-amp voltage comparator. A combination of voltage comparator circuits and digital circuits are used inside these chips. Although several devices are on the market, the most common and best known is the type 555 device. The 555 is now made by a number of different semiconductor manufacturers, but was originated by Signetics, Inc. in 1970. Today the 555 remains one of the most widespread IC devices on the market, rivaling even some general purpose operational amplifiers in numbers.

The original Signetics products included the SE-555, which operated over a temperature range of −55 to +125°C, and the NE-555, which operated over the range 0 to +70°C. Several different designations are now commonly used for the 555 made by other makers, including simply '555' and 'LM-555', or some variant numbering. A dual 555-class timer is also marketed under the number '556'. There is also a low-power CMOS version of the 555 marketed as the LMC-555.

The 555 is a multi-purpose chip that will operate at DC power supply potentials from +5 Vdc to +18 Vdc. The temperature stability of these devices is on the order of 50 PPM/°C (i.e. 0.005%/°C). The output of the 555 can either sink or source up to 200 mA of current. It is compatible with TTL devices (when the 555 is operated from +5 Vdc power supply), CMOS devices, operational amplifiers, other linear IC devices, transistors and most classes of solid-state devices. The 555 will also operate with most passive electronic components.

Several factors contribute to the popularity of the 555 device. Besides the versatile nature of the device, it is well-behaved in the sense that operation is straightforward and circuit designs are generally simple. Like the general purpose operational amplifier, the 555 usually works in a predictable manner, according to the standard published equations.

The 555 operates in two different modes: monostable (one-shot) and astable (free-running). Figure 4-1A shows the astable mode output from pin no. 3 of the 555. The waveform is a series of squarewaves that can be varied in duty cycle over the range 50 to 99.9%, and in frequency from less than 0.1 Hz to more than 100 kHz. Monostable operation (Fig. 4-1B) requires a trigger pulse applied to pin no. 2 of the 555. The trigger must drop from a level $>2(V+)/3$ down to $<(V+)/3$. Output pulse durations from microseconds up to hours are possible. The principal constraint on longer operation is the leakage resistance of the capacitor used in the external timing circuit.

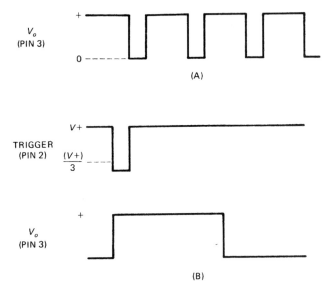

FIGURE 4-1 (A) 555 IC timer astable multivibrator output; (B) monostable multivibrator trigger and output waveforms.

4-3 PINOUTS AND INTERNAL CIRCUITS OF THE 555 IC TIMER

The package for the 555 device is shown in Fig. 4-2. Most 555s are sold in the eight-pin miniDIP package as shown, although some are still found in the eight-pin metal can IC package. The latter are mostly the military specification temperature range SE-555 series. The pinouts are the same on both miniDIP and metal can versions. The internal circuitry is shown in block form in Fig. 4-3. The following stages are found: two voltage comparators ($COMP1$ and $COMP2$), a reset–set (RS) control flip-flop (which can be reset from outside the chip through pin no. 4), an inverting output amplifier ($A1$) and a discharge transistor ($Q1$). The bias levels of the two comparators are determined by a resistor voltage divider (R_a, R_b and R_c) between $V+$ and ground. The inverting input of $COMP1$ is set to $2(V+)/3$, and the noninverting input of $COMP2$ is set to $(V+)/3$.

Figures 4-3 and 4-4 show the pinouts of the 555. In the descriptions below the term HIGH implies a level $>2(V+)/3$, and LOW implies a grounded condition ($V = 0$), unless otherwise specified in the discussion. These pins serve the following functions:

Ground (Pin No. 1). This pin serves as the common reference point for all signals and voltages in the 555 circuit both internal and external to the chip.

Trigger (Pin No. 2). The trigger pin is normally held at a potential $>2(V+)/3$. In this state the 555 output (pin no. 3) is LOW. If the trigger

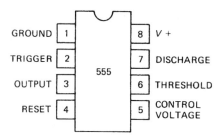

FIGURE 4-2 555 package and pinouts.

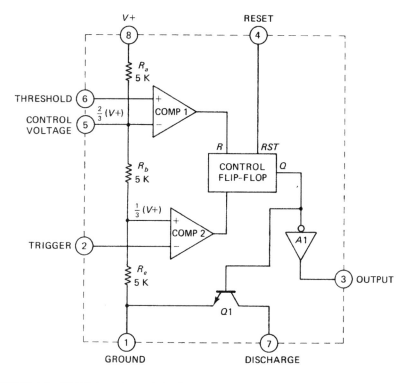

FIGURE 4-3 555 IC timer internal circuit.

pin is brought LOW to a potential $<(V+)/3$, then the output (pin no. 3) abruptly switches to the HIGH state. The output remains HIGH as long as pin no. 2 is LOW, but the output does not necessarily revert back to LOW immediately after pin no. 2 is brought HIGH again (see operation of the Threshold input below).

Output (Pin No. 3). The output pin of the 555 is capable of either sinking or sourcing current up to 200 mA. This operation is in contrast to other IC

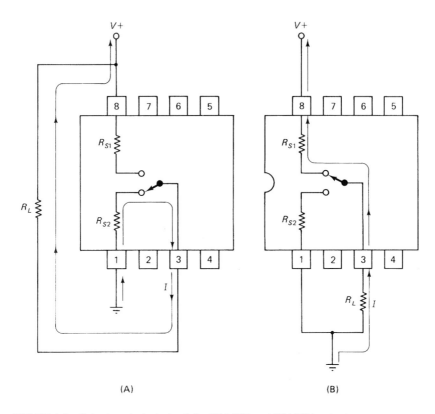

FIGURE 4-4 Output equivalent circuit for (A) LOW and (B) HIGH outputs.

devices in which the outputs of various devices will either sink or source current, but not both. Whether the 555 output operates as a sink or a source depends on the configuration of the external load. Figure 4-4 shows both types of operation.

In Fig. 4-4A the external load R_L is connected between the 555 output and $V+$. Current only flows in the load when pin no. 3 is LOW. In that condition the external load is grounded through pin no. 1 and a small internal source resistance, R_{s1}. In this configuration the 555 output is a current sink.

The operation depicted in Fig. 4-4B is for the case where the load is connected between pin no. 3 of the 555 and ground. When the output is LOW the load current is zero. But when the output is HIGH, however, the load is connected to $V+$ through a small internal resistance R_{s2} and pin no. 8. In this configuration the output serves as a current source.

Reset (Pin No. 4). The reset pin is connected to a preset input of the 555 internal control flip-flop. When a LOW is applied to pin no. 4 the output of the 555 (pin no. 3) switches immediately to a LOW state. In normal operation it is common practice to connect pin no. 4 to $V+$ in order to prevent false resets from noise impulses.

Control Voltage (Pin No. 5). This pin normally rests at a potential of $2(V+)/3$ due to an internal resistive voltage divider (see R_a through R_c in Fig. 4-3). Applying an external voltage to this pin, or connecting a resistor to ground, will change the duty cycle of the output signal. If not used, then pin no. 5 should be decoupled to ground through a 0.01 μF to 0.1 μF capacitor.

Threshold (Pin No. 6). This pin is connected to the noninverting input (+IN) of comparator $COMP1$, and is used to monitor the voltage across the capacitor in the external RC timing network. If pin no. 6 is at a potential of $<2(V+)/3$, then the output of the control flip-flop is LOW, and the output (pin no. 3) is HIGH. Alternatively, when the voltage on pin no. 6 is $\geq 2(V+)/3$, then the output of $COMP1$ is HIGH and chip output (pin no. 3) is LOW.

Discharge (Pin No. 7). The discharge pin is connected to the collector of NPN transistor $Q1$, and the emitter of $Q1$ is connected to the ground pin (no. 1). The base of $Q1$ is connected to the NOT-Q output of the control flip-flop. When the 555 output is HIGH, the NOT-Q output of the control flip-flop is LOW, so $Q1$ is turned off. The c–e resistance of $Q1$ is very high under this condition, so does not appreciably affect the external circuitry. But when the control flip-flop NOT-Q output is HIGH, however, the 555 output is LOW and $Q1$ is biased hard on. The c–e path is in saturation, so the c–e resistance is very low. Pin no. 7 is effectively grounded under this condition.

$V+$ Power Supply (Pin No. 8). The DC power supply is connected between ground (pin no. 1) and pin no. 8, with pin no. 8 being positive. In good practice a 0.1 μF to 10 μF decoupling capacitor will normally be used between pin no. 8 and ground.

4-4 MONOSTABLE OPERATION OF THE 555 IC TIMER

A monostable multivibrator (MMV), also called the one-shot, circuit produces a single output pulse of fixed duration when triggered by an input pulse. The output of the one-shot will snap HIGH following the trigger pulse, and will remain HIGH for a fixed, predetermined duration. When this time expires the one-shot is timed-out so snaps LOW again. The output of the one-shot will remain LOW indefinitely unless another trigger pulse is applied to the circuit. The 555 can be operated as a monostable multivibrator by suitable connection of the external circuit.

Figure 4-5 shows the operation of the 555 as a monostable multivibrator. In order to make the operation of the circuit easier to understand, Fig. 4-5A shows the internal circuitry as well as the external circuitry; Fig. 4-5B shows the timing diagram for this circuit; and Fig. 4-5C shows the same circuit in the more conventional schematic diagram format.

The two internal comparators are biased to certain potential levels by a series voltage divider consisting of resistors R_a, R_b and R_c. The inverting input of voltage comparator $COMP1$ is biased to $2(V+)/3$, while the noninverting input of $COMP2$ is biased to $(V+)/3$. It is these levels that govern the operation of the 555 device in whichever mode is selected. An external timing network $(R1C1)$ is connected between $V+$ and the noninverting input of $COMP1$ via pin no. 6. Also connected to pin no. 6 is 555 pin no. 7, which has the effect of connecting the transistor across capacitor $C1$. If the transistor is turned on, then the capacitor looks into a very low resistance short circuit through the c–e path of the transistor.

When power is initially applied to the 555, the voltage at the inverting input of $COMP1$ will go immediately to $2(V+)/3$ and the noninverting input of $COMP2$ will go to $(V+)/3$. The control flip-flop is in the reset condition, so the NOT-Q output is HIGH. Because this flip-flop is connected to output pin no. 3 through an inverting amplifier ($A1$), the output is LOW at this point. Also, because NOT-Q is HIGH, transistor $Q1$ is biased into saturation,

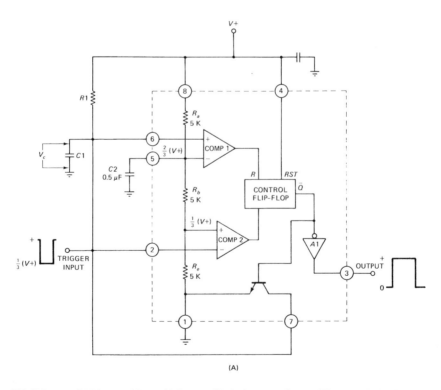

(A)

FIGURE 4-5 (A) Monostable multivibrator; (B) timing waveforms; (C) external circuitry.

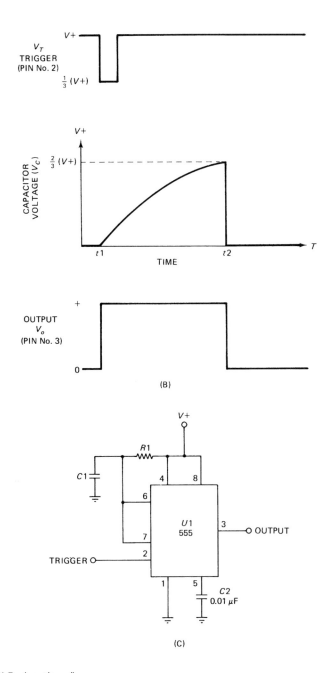

FIGURE 4-5 (continued)

INTEGRATED CIRCUIT TIMERS **121**

creating a short circuit to ground across external timing capacitor $C1$. The capacitor remains discharged in this condition ($V_c = 0$). If a trigger pulse is applied to pin no. 2 of the 555, and if that pulse drops to a voltage that is $<(V+)/3$, as shown in Fig. 4-5B, then comparator $COMP2$ sees a situation where the inverting input is less positive than the noninverting input, so the output of $COMP2$ snaps HIGH. This action sets the control flip-flop, forcing the NOT-Q output LOW, and therefore the 555 output HIGH. The LOW at the output of the control flip-flop also means that transistor $Q1$ is now unbiased, so the short across the external capacitor is removed. The voltage across $C1$ begins to rise (see Figs 4-5B and 4-5D). The voltage will continue to rise until it reaches $2(V+)/3$, at which time comparator $COMP1$ will snap HIGH causing the flip-flop to reset. When the flip-flop resets, its NOT-Q output drops LOW again, terminating the output pulse, and returning the capacitor voltage to zero. The 555 will remain in this state until another trigger pulse is received.

The timing equation for the 555 can be derived in exactly the same manner as the equations used with the operational amplifier MMV circuits. The basic equation relates the time required for a capacitor voltage to rise from a starting point (V_{C1}) to an end point (V_{C2}) with a given RC time constant:

$$T = -RC \ln\left[\frac{V - V_{C2}}{V - V_{C1}}\right] \quad (4\text{-}1)$$

In the 555 timer the voltage source is $V+$, the starting voltage is zero, and the trip-point voltage for comparator $COMP1$ is $2(V+)/3$. Equation (4-1) can therefore be rewritten as:

$$T = -R1C1 \ln\left[\frac{V - V_{C2}}{V - V_{C1}}\right] \quad (4\text{-}2)$$

$$T = -R1C1 \ln\left[\frac{(V+) - 2(V+)/3}{V+}\right] \quad (4\text{-}3)$$

$$T = -R1C1 \ln(1 - 0.667) \quad (4\text{-}4)$$

$$T = -R1C1 \ln(0.333) \quad (4\text{-}5)$$

$$T = 1.1 R1C1 \quad (4\text{-}6)$$

4-4.1 Input triggering methods for the 555 MMV circuit

The 555 MMV circuit triggers by bringing pin no. 2 from a positive voltage down to a level $<(V+)/3$. Triggering can be accomplished by applying a pulse from an external signal source, or through other means. Figure 4-6 shows the circuit for a simple pushbutton switch trigger circuit. A pull-up resistor ($R2$) is connected between pin no. 2 and $V+$. If normally-open (NO) pushbutton switch $S1$ is open, then the trigger input is held at a potential very

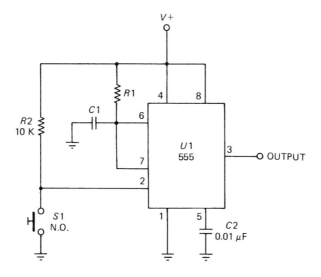

FIGURE 4-6 Switch triggered monostable multivibrator circuit.

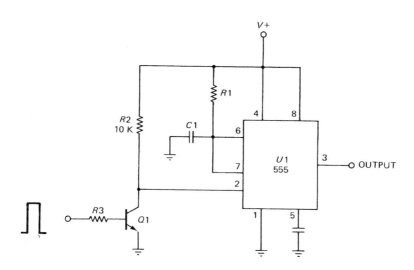

FIGURE 4-7 Positive triggering monostable multivibrator.

close to $V+$. But when $S1$ is closed, pin no. 2 is brought LOW to ground potential. Because pin no. 2 is now at a potential less than $(V+)/3$ the 555 MMV will trigger. This circuit can be used for contact debouncing.

A circuit for inverting the trigger pulse applied to the 555 is shown in Fig. 4-7. In this circuit an NPN bipolar transistor is used in the common emitter mode to inverting the pulse. Again, a pull-up resistor is used to keep pin no. 2 at $V+$ when the transistor is turned off. But when the positive polarity trigger pulse is received at the base of transistor $Q1$, the transistor saturates and this forces the collector (and pin no. 2 of the 555) to near ground potential.

Figure 4-8 shows two AC-coupled versions of the trigger circuit. In these circuits a pull-up resistor keeps pin no. 2 normally at $V+$. But when a pulse is applied to the input end of capacitor $C3$, a differentiated version of the pulse is created at the trigger input of the 555. Diode $D1$ clips the positive going spike to 0.6 or 0.7 volts, passing only the negative-going pulse to the 555. If the negative-going spike can counteract the positive bias provided by $R2$ sufficiently to force the voltage lower than $(V+)/3$, then the 555 will trigger. A pushbutton switch version of this same circuit is shown in Fig. 4-8B.

A touchplate trigger circuit is shown in Fig. 4-9A. The pull-up resistor $R2$ has a very high value (22 megohms shown here). The touchplate consists of

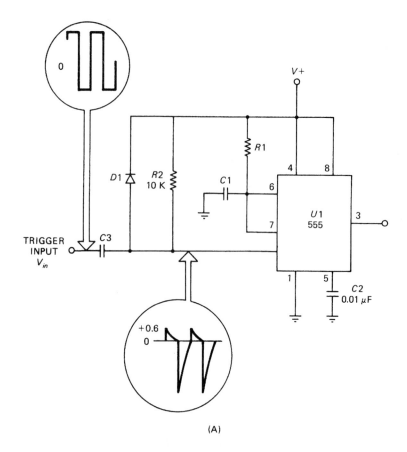

FIGURE 4-8 (A) pulse triggering; (B) switch triggering.

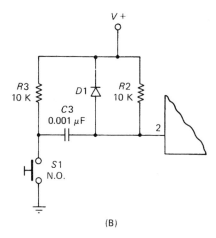

(B)

FIGURE 4-8 (continued)

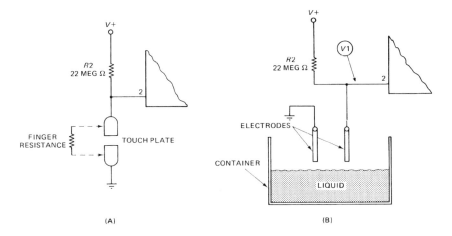

FIGURE 4-9 (A) Touch-plate switching circuit; (B) liquid level detector.

a pair of closely spaced electrodes. As long as there is no external resistance between the two halves of the touchplate, the trigger input of the 555 remains at $V+$. But when a resistance is connected across the touchplate, the voltage ($V1$) drops to a very low value. If the average finger resistance is about 20 kohms, the voltage drops to:

$$V1 = \frac{(V+)(20 \text{ k}\Omega)}{(R2 + 20 \text{ k}\Omega)} \qquad (4\text{-}7)$$

which is, when $R2 = 22$ megohms, to $0.0009\ (V+)$ — which is certainly less than $(V+)/3$.

INTEGRATED CIRCUIT TIMERS **125**

The same concept is used in the liquid level detector shown in Fig. 4-9B. Once again a 22 megohm pull-up resistor is used to keep pin no. 2 at $V+$ under normal operation. When the liquid level rises sufficiently to short out the electrodes, however, the voltage on pin no. 2 ($V1$) drops to a very low level, forcing the 555 to trigger.

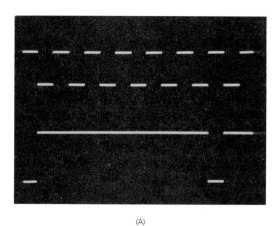

(A)

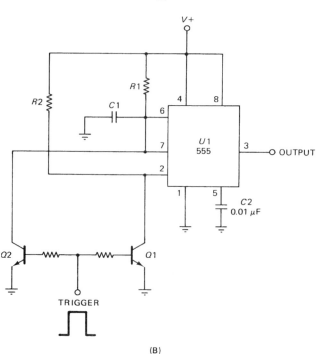

(B)

FIGURE 4-10 (A) Retriggerable monostable multivibrator waveforms; (B) retriggerable multivibrator circuit.

4-4.2 Retriggerable operation of the 555 MMV circuit

The 555 is a nonretriggerable monostable multivibrator. If additional trigger pulses are received prior to the time-out of the output pulse (see Fig. 4-10A), then the additional pulses have no effect on the output. But the first pulse after time-out occurs will cause the output to again snap HIGH. In Fig. 4-10A the trigger input signal is a squarewave signal, so the output of the 555 is LOW only for the duration of one trigger pulse before snapping HIGH again.

The circuit in Fig. 4-10B will permit retriggering of the 555 device. An external NPN transistor ($Q2$ in Fig. 4-10B) is connected with its c-e path across timing capacitor $C1$. In this sense it mimmicks the internal discharge transistor seen earlier. A second transistor, $Q1$, is connected to the trigger input of the 555 in a manner similar to Fig. 4-7 (discussed earlier). The bases of the transistors form the trigger input. When a positive pulse is applied to the combined trigger line both transistors become saturated. Any charge in $C1$ is immediately discharged, and pin no. 2 of the 555 is triggered by the collector of $Q1$ being dropped to less than $(V+)/3$.

As long as no further trigger pulses are received, this circuit behaves like any other 555 MMV circuit. But if a trigger pulse is received prior to the time-out defined by Eq. (4-6), the transistors are forward biased once again. $Q1$ retriggers the 555, while $Q2$ dumps the charge built up in the capacitor. Thus, the 555 retriggers.

4-5 APPLICATIONS FOR THE 555 ONE-SHOT CIRCUIT

The MMV is a one-shot circuit that produces a single output pulse for every trigger input pulse, except for those that fall inside the output pulse and any associated refractory period. There are numerous potential applications for these circuits, of which a few are presented in this section.

4-5.1 Missing pulse detector

A *missing pulse detector* circuit remains dormant as long as a series of trigger pulses are received, but will produce an output pulse when an expected pulse is missing. These circuits are used in a variety of applications including alarms. For example, in a bottling plant softdrink cans are packaged into six-packs. As each can passes a photocell a pulse is generated to the input of a missing pulse detector. If a pulse is not received, however, the machine knows that the count is one can short, so issues an alarm or corrective action. Similarly in a wildlife photography system. An infrared light emitting diode (LED) is modulated or chopped with a pulse waveform. As long as the pulse is received at the sensor, the circuit is dormant. But if an animal passes through the IR beam even briefly a missing pulse detector will sense its presence and issue an output that fires a camera flashgun and electrical shutter control.

Figure 4-11A shows the circuit for a missing pulse detector based on the 555 IC timer, while Fig. 4-11B shows the timing waveforms. This circuit is the standard 555 MMV, except that a discharge transistor is shunted across capacitor C1. When a pulse is applied to the input it will trigger the 555 and turn on Q1, causing the capacitor to discharge. After the first input pulse the output of the 555 snaps HIGH and remains HIGH until a missing pulse is detected.

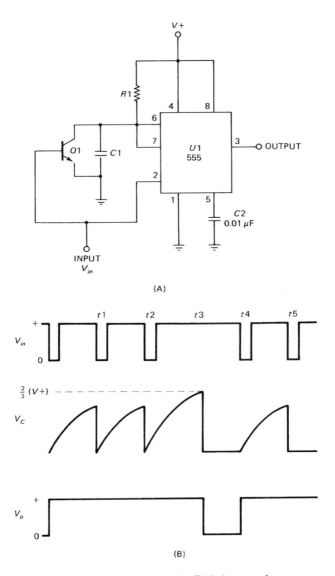

FIGURE 4-11 Missing pulse detector: (A) circuit; (B) timing waveforms.

Circuit action can be seen in Fig. 4-11B. At times $t1$ and $t2$ input pulses are received. As long as $(t2 - t1)$ is less than the time required for $C1$ to charge up to $2(V+)/3$ the 555 will never time-out. But if a pulse is missing, as at $t3$, the capacitor voltage continues to rise to the critical $2(V+)/3$ threshold value. When V_c reaches this point the 555 will time-out forcing its output LOW. The output remains LOW until a subsequent input pulse is received ($t4$), at which time $Q1$ turns on again and forces the capacitor to discharge. The cycle can then continue as before.

4-5.2 Pulse position circuit

A *pulse positioner* is a circuit that will allow adjustment of the timing of a pulse to coincide with some external event. For example, in some instrumentation circuits a short pulse must be positioned to a certain point on a sinewave (e.g. the peak). The pulse positioner could be triggered from the zero-crossing of the sinewave, and then adjusted to place the output pulse where it is needed.

Figure 4-12A shows the concept of pulse positioning using two one-shot circuits, labeled $OS1$ and $OS2$. The circuit is shown in Fig. 4-12B. The re-positioned pulse is not actually the original pulse, but rather it is a recreated pulse with similar characteristics. The input pulse is used to trigger $OS1$. The duration of this one-shot circuit is fixed to the delay required of the re-positioned pulse. If the delay must be variable, then resistor $R1$ is made variable. When $OS1$ times-out it will trigger $OS2$. The output pulse of $OS2$ is set to the parameters of the original input pulse. An inverter circuit is used to make the output of $OS2$ have the same polarity as the trigger pulse at the input of $OS1$. To an outside observer the pulse appears to have been re-positioned, although in fact it was merely recreated at time T (the delay period in Fig. 4-12B).

4-5.3 Tachometry

The word *tachometry* is used to designate the measurement of a repetition rate. In the automotive tachometer, for example, the instrument counts the pulses produced by the ignition coil to measure the engine speed in RPM. In medical instruments it is often necessary to measure factors such as heart or respiration rate electronically using tachometry circuits. A heart rate meter (*cardiotachometer*) measures the heart rate in beats per minute (BPM), while the respiration meter (*pneumotachometer*) measures breathing rate in breaths per second.

There is a certain commonality among non-digital tachometer circuits. It doesn't matter whether the rate is audio or sub-audio, or even above the audio rate, the basic circuit design is the same.

Figure 4-13 shows the basic tachometer circuit in block diagram form. Not all of the stages will be present in all circuits, but some of them are basic to

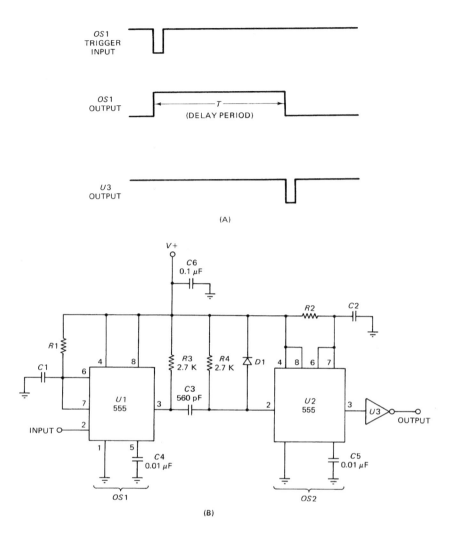

FIGURE 4-12 (A) stretched pulse waveform; (B) pulse stetcher circuit.

FIGURE 4-13 Block diagram of tachometer circuit.

the problem so are universally found. The AC amplifier and Schmitt trigger will be used only when needed. These stages are used for input signal conditioning, so are used only where such conditioning is needed. The one-shot circuit and the Miller integrator are basic to the design, however, so are used for all such circuits.

The idea is to convert a frequency or repetition rate to an analog voltage. This is done by first converting the signal to pulse form. The AC input amplifier is used only if it is necessary to scale the input signal to a level where it will drive a Schmitt trigger or other squaring circuit. The purpose of the following stage is to produce a squarewave output signal at the same frequency of as the input signal.

The purpose of the stages in Fig. 4-13 is to produce a DC voltage output that is proportional to the input frequency or pulse repetition rate. The integrator is designed to produce an output voltage that is the time-average of the input signal. That is, the integrator output is proportional to the area under the input signal. The job of the tachometer designer is to create a situation in which the only variable is the frequency or repetition rate of the input signal. Variation in other factors obscures the results.

The output pulse of a one-shot circuit has a constant amplitude and constant duration. The area under the pulse is the product of the amplitude and duration, so from pulse to pulse the area does not change. If the one-shot is constantly retriggered by the input signal, the total area under the resultant pulse train is a function of only the number of pulses. Therefore, the time-average of the integrator output will be a DC voltage that is proportional to the input frequency.

Figure 4-14 shows a practical application of the tachometer principle. The circuit was used to demodulate the audio frequency modulated signal from an instrumentation telemetry set. A similar circuit (but not based on the 555) was once popular as a coilless FM detector in communications and broadcast receivers. These *pulse counting detectors* operated at 10.7 MHz (a commonly used FM IF frequency in receivers). The circuit shown in Fig. 4-14 was used to demodulate a human electrocardiograph (ECG) signal transmitted over telephone lines. The ECG is an analog voltage waveform, and was used to frequency modulate an audio voltage controlled oscillator (VCO) at the transmit end. Normally, the ECG has too low a Fourier frequency content (0.05 Hz to 100 Hz) to pass over the restricted passband of the telephone lines (300 Hz to 3000 Hz). But when used to frequency modulate a 1500 Hz carrier, however, the signal passed easily over telephone circuits.

The circuit for the demodulator circuit is shown in Fig. 4-14A. The input waveshaping function is performed by an LM-311 voltage comparator. The job of the LM-311 is to square the 200 mV peak-to-peak sinewave input signal so that it is capable of triggering the 555 ($U2$). In this mode the LM-311 is operating basically as a zero-crossing detector circuit.

The output of the 555 is a pulse train that has constant amplitude and duration. These pulses vary only in repetition rate, which is the same as the frequency of the input signal. The 555 output pulses are integrated in a passive *RC* integrator ($R5-R7/C4-C6$). The output of the integrator is a DC voltage that is a linear function of input frequency (see Fig. 4-14B). This DC voltage can be scaled, if necessary, to any desired level.

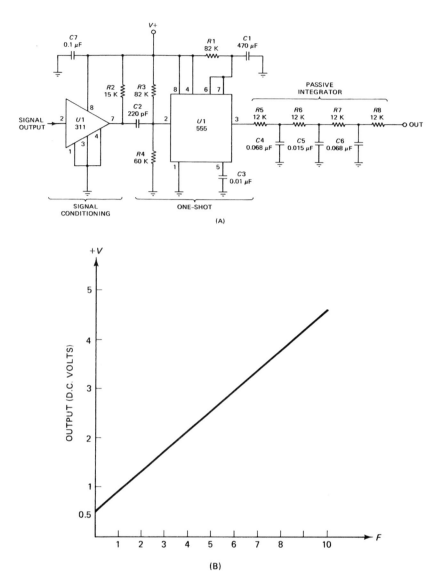

FIGURE 4-14 (A) 555-based tachometer circuit; (B) output transfer function.

A related circuit is shown in Fig. 4-15. This 555-based tachometer is used to measure audio frequency over three ranges: DC to 50 Hz, DC to 500 Hz and DC to 5000 Hz. The circuit uses the same form of input signal conditioning as the previous circuit, and uses a 555 as the one-shot circuit. The integration function is taken up by the combination of RC network $R4/C4$ and the mechanical inertia of the meter ($M1$) movement.

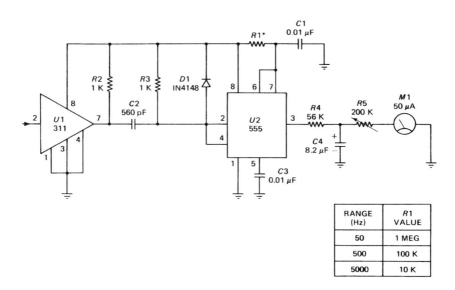

FIGURE 4-15 555-based analogue frequency meter.

4-6 ASTABLE OPERATION OF THE 555 IC TIMER

An astable multivibrator (AMV) is a free-running circuit that produces a squarewave output signal. The 555 can be connected to produce a variable duty cycle AMV circuit (Fig. 4-16). A version of the circuit showing the internal stages of the 555 is shown in Fig. 4-16A, while the circuit as it normally appears in schematic drawings is shown in Fig. 4-16B. The factor that makes this circuit an AMV is that the threshold and trigger pins (6 and 2) are connected together, forcing the circuit to be self-retriggering.

Under initial conditions at turn-on the voltage across timing capacitor $C1$ is zero, while the biases on $COMP1$ and $COMP2$ are (as usual) set to $2(V+)/3$ and $(V+)/3$, respectively, by the internal resistor voltage divider (R_a, R_b and R_c). The output of the 555 is HIGH under this condition, so $C1$ begins to charge through the combined resistance $[R1 + R2]$. On discharge, however, transistor $Q1$ shorts the junction of $R1$ and $R2$ to ground, so the capacitor discharges through only $R2$. The result is the waveform shown in Fig. 4-16C. The time that the output is HIGH is $t1$, while the LOW time is $t2$. The period (T) of the output squarewave is the sum of these two durations: $T = (t1 = t2)$.

As with all similar RC-timed circuits the equation that sets oscillating frequency is determined from Eq. (4-8) below:

$$T = -R1C1 \ln \left[\frac{V - V_{C2}}{V - V_{C2}} \right] \tag{4-8}$$

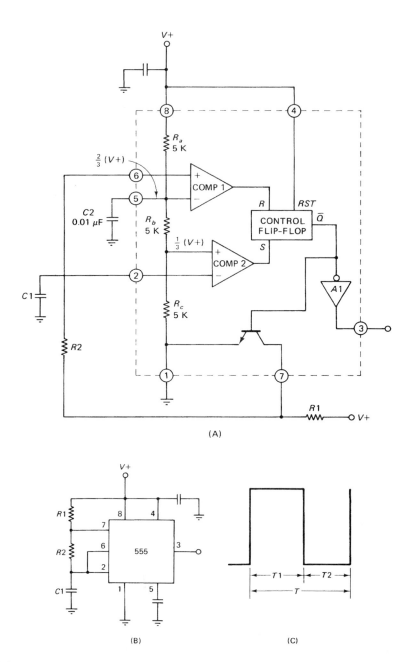

FIGURE 4-16 (A) Astable multivibrator circuit showing 555 internal circuits; (B) external circuits; (C) output waveform; (D) output and capacitor voltage waveforms.

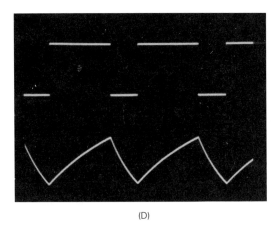

(D)

FIGURE 4-16 *(continued)*

For the case where the output is HIGH ($t2$ in Fig. 4-16C), the resistance R is $[R1 + R2]$, and the capacitance (C) is $C1$. Because of the internal biases of the voltage comparator stages of the 555, the capacitor will charge from $(V+)/3$ to $2(V+)/3$, and then discharge back to $(V+)/3$ on each cycle. Thus, Eq. (4-8) can be rewritten:

$$t1 = -(R1 + R2)C1 \ln \left[\frac{(V+) - 2(V+)/3}{(V+) - (V+)/3} \right] \tag{4-9}$$

or, once the algebra is done:

$$t1 = 0.695(R1 + R2)C1 \tag{4-10}$$

By similar argument it can be shown that

$$t2 = 0.695 \, R2C1 \tag{4-11}$$

For the total period T:

$$T = t1 + t2 \tag{4-12}$$

$$T = [0.695(R1 + R2)C1] + [0.695 \, R2C1]$$

$$T = 0.695(R1 + 2R2)C1 \tag{4-13}$$

Equation (4-13) defines the *period* of the output squarewave. In order to find the *frequency of oscillation* take the reciprocal of Eq. (4-13):

$$F = \frac{1}{T} \tag{4-14}$$

$$F = \frac{1.44}{(R1 + 2R2)C1} \tag{4-15}$$

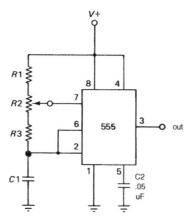

FIGURE 4-17 Variable frequency astable multivibrator.

4-6.1 Duty cycle of 555 astable multivibrator

Time segments $t1$ and $t2$ are not equal in most cases, so the charge and discharge times for capacitor $C1$ are also not equal (see Fig. 4-16D). The duty cycle of the output signal is the ratio of the HIGH period to the total period $(t1/T)$. Expressed as a percentage:

$$\%DC = \frac{R1 + R2}{R1 + 2R2} \tag{4-16}$$

Various methods are used for varying the duty cycle. First, a voltage can be applied to pin no. 5 (*Control Voltage*). Second, a resistance can be connected from pin no. 5 to ground. Both of these tactics have the effect of altering the internal bias voltages applied to the comparator. Alternatively, one can also divide the external resistances $R1$ and $R2$ into three values. Figure 4-17 shows a variable duty factor 555 AMV that uses a potentiometer ($R2$) to vary the ratio of the charge and discharge resistances.

4-6.2 Synchronized operation of 555 astable multivibrator

A synchronized AMV operates in a manner in which its operating frequency is locked to an external frequency. The horizontal and vertical deflection oscillators in a television receiver operate in this manner because they are locked to the sync pulses transmitted by the TV broadcast station.

A method for locking in the oscillating frequency of the 555 is shown in Fig. 4-18. In this circuit a 7400 TTL NAND gate is used to sample both the 555 AMV output signal and the input sync signal. The properties of the NAND gate are these:

1. If either input is LOW, then the output is HIGH.
2. Both inputs must be HIGH for the output to be LOW.

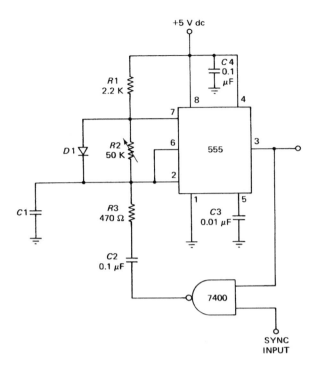

FIGURE 4-18 Synchronized 555 astable multivibrator.

Because the output of the NAND gate is applied to the timing circuit of the 555 (through waveshaping network $R3/C2$), it will affect the relative timing of the circuit. This circuit is analogous to a mechanical pendulum oscillator which has an external non-resonant forcing frequency applied. The circuit will lock to the new frequency if it is reasonably close to the natural oscillating frequency, or an integer harmonic of the natural frequency. Students interested in modern chaos theory might want to investigate the behavior of this and similar circuits at sync frequencies away from the natural frequency or its harmonics and subharmonics.

4-6.3 A 555 Sawtooth generator circuit

A sawtooth waveform (Fig. 4-19A) rises linearly to some value, and the drops abruptly back to the initial conditions.

The circuit for a 555-based sawtooth generator (Fig. 4-19B) is simple, and is based on the 555 timer IC. The basic circuit is the monostable multivibrator configuration of the 555, in which one of the timing resistors is replaced with a transistor operated as a current source ($Q1$). Almost any audio small-signal PNP silicon replacement transistor can be used, although for this test the 2N3906 device was used. The zener diode is a 5.6 Vdc unit. Note that the

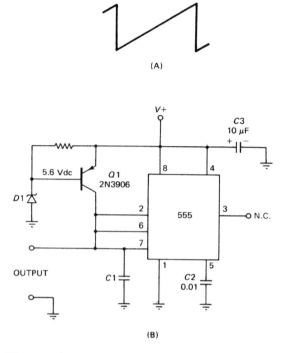

FIGURE 4-19 555 sawtooth generator circuit.

output is taken from pins 6–7, rather than the regular chip output, pin no. 3, which is not used.

The circuit as shown is a one-shot multivibrator. Triggering occurs in the 555 when pin no. 2 is brought to a potential less than 2/3 the supply potential. When a pulse is applied to pin no. 2 through differentiating network $R1C1$, the device will trigger because the negative-going slope meets the triggering criteria. To make an astable sawtooth multivibrator drive the input of this circuit with either a squarewave or pulse train that produces at least one pulse for each required sawtooth. Being a nonretriggerable monostable multivibrator the circuit of Fig. 4-19B will ignore subsequent trigger pulses during the one-shot's 'refractory' period.

4-7 XR-2240 IC TIMER

The XR-2240 (also designated 8240) IC timer (Fig. 4-20) is based on the same circuit concept as the 555 device; i.e. a window comparator that sets and resets a control flip-flop. The timebase circuit must receive its power from outside the chip, so in normal operation a 20 kohm resistor is connected between the regulator output (pin no. 15) and timebase output (pin no. 14). This feature allows external timebase circuits to be used.

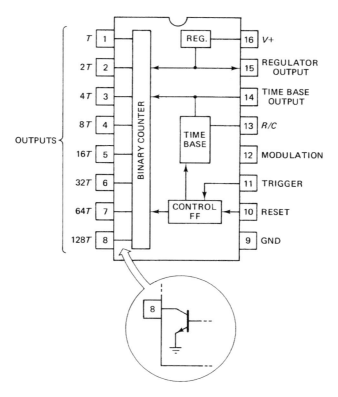

FIGURE 4-20 XR-2240 IC circuit.

The timebase section of the XR-2240 is basically the 555-style timer modified to remove the constant 1.1 from the timing equation (Eq. (4-6)). The reason for the constant in the 555 is that the resistors in the internal voltage divider that biases the comparators are equal. But in the XR-2240 non-equal resistances are used, making the reference levels $0.27(V+)$ and $0.73(V+)$ instead of $0.33(V+)$ and $0.67(V+)$ as used in the 555. This change was made in order to simplify the timing equation to:

$$T = RC \qquad (4\text{-}17)$$

The values of the timing components are 1 kohm $\leq R \leq$ 10 megohms, and 0.05 µF $\leq C \leq$ 1000 µF. The XR-2240 is packaged in a 16-pin DIP case, and will operate over the range +4.5 Vdc to +18 Vdc. The XR-2240 differs from the 555 in that the trigger and reset pulses are positive-going rather than negative-going. The minimum amplitude of the trigger and reset pulses is approximately two PN-junction voltage drops, or about 1.4 Vdc. As a practical matter, however, a minimum 3 Vdc level is recommended in order to guard against trigger failures caused by negative noise impulses.

There are other differences between the 555 device and the XR-2240. The XR-2240 contains an eight-bit binary counter with open-collector NPN transistor outputs (see inset to Fig. 4-20). If these outputs are wired in the logical-OR configuration, then times from $1T$ to $255T$ can be programmed, where T is defined by Eq. (4-17) above. Each output is connected to $V+$ through a 10 kohms pull-up resistor. The use of the open-collector configuration makes the output active-LOW. The combination of the large range of timer RC components and an eight-bit binary counter makes it possible to create very long duration timer circuits that are essentially as stable as the RC network.

It is also possible to use the binary outputs independently of each other if each is supplied with its own 10 kohm pull-up resistor. Examples of such applications include binary address sequences for digital circuits, and oscillators with base-2 related synchronized output frequencies.

Other, less common, versions of the timer include the XR-2250 (or 8250), which offers *binary coded decimal* (BCD) outputs instead of binary, and the XR-2260, which uses BCD outputs but limits the most significant digit to 6.

Figure 4-21 shows the circuit for both the monostable and astable multivibrator configurations for the XR-2240. One interesting design feature of the XR-2240 is that the sole difference between the monostable and astable modes is a single feedback resistor ($R3$) that automatically re-triggers the XR-2240 at the end of each output cycle.

The timer is set into operation by applying a positive-going pulse to the trigger input (pin no. 11) This pulse is routed to the control logic, and performs several jobs simultaneously: resetting the binary counter to 00000000_2, driving all outputs HIGH, and enabling the timebase circuit. As was true with the 555 timer, the XR-2240 works by charging capacitor $C1$ through resistor $R1$ from positive voltage source $V+$.

One purpose for the open-collector configuration is that different multiples of the basic timer duration can be programmed by connecting the required outputs together in a wired-OR configuration. For example, if a 57 second timer is needed it is possible to use a 1 megohm resistor and a 1 µF capacitor ($T = 1$ second), and then connect together the $1T$, $8T$, $16T$ and $32T$ outputs: $(1 + 8 + 16 + 32)T = 57T$. In that circuit pins 1, 4, 5 and 6 will be connected together, and to $V+$ through a single 10 kohm resistor. The output will remain LOW for 57 seconds following triggering, and then returns to the HIGH state.

The XR-2240 is considered by some to be more flexible than the 555 because the duration (monostable mode) or period (astable mode) can be set by either the RC timing network or through selection of which outputs are wired together at the output of the circuit.

Synchronization to an external timebase, or modulation of the pulse width, is possible by manipulation of the *modulation* input (pin no. 12). In normal operation, the modulation input is bypassed to ground through a 0.01 µF

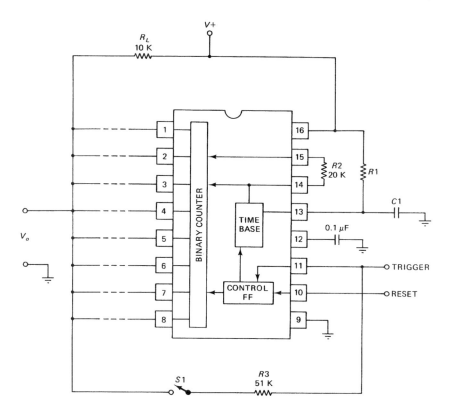

FIGURE 4-21 XR-2240 monostable multivibrator circuit.

capacitor so that noise signals will not disrupt operation of the device. A voltage applied to pin no. 12 will modulate the pulse width of the timebase output signal. This modulating voltage should be between +2 and +5 volts for a change factor of 0.4 to 2.25.

Synchronization of the XR-2240 to an external timebase is accomplished through a series RC network consisting of a 5.1 kohm resistor and a 0.1 μF capacitor (see Fig. 4-22) connected to pin no. 12. This network differentiates the input pulse. The synchronization pulse should have an amplitude of at least +3 volts, and a period of $0.3T$ to $0.8T$. Another method of synchronization is to use an external timebase connected directly to pin no. 14.

4-7.1 Very long duration timers

The long-duration timer presents special problems to the electronic circuit designer. The drift and other errors of some components tend to accumulate, so the total error becomes large over longer time periods. In fact, whenever the errors are a function of time the long duration timer will suffer markedly.

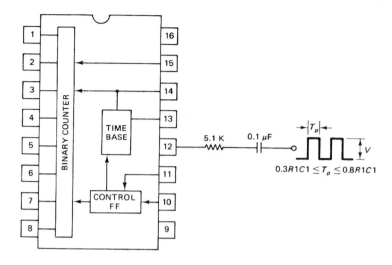

FIGURE 4-22 Synchronizing circuit.

Any long duration RC network, for example, will suffer from several time related problems.

A viable alternative is to use an RC timer. But most high value capacitors are electrolytic types, and as a result have a tolerance of -20 to $+100\%$ of the label capacitance. In addition, some capacitors show capacitance variation with both time and applied voltage. In addition, many large-value capacitors exhibit considerable shunt resistance that must be considered in timing networks. Because of these problems the 555 is limited to practical durations of about 100 seconds if precision is a consideration.

The timebase section of the XR-2240 timer suffers from the same problems as does the 555, but the RC values needed for very long duration monopulse operation are much lower than the 555 because of the built-in binary counter. If the outputs of the XR-2240 are wired-OR together, then times from $1RC$ to $255RC$ are possible. For example, with an RC time constant of 10 seconds, a single XR-2240 can be used for up to 2550-seconds (42.5 minutes). Greater duration times can be accommodated by cascading two or more XR-2240 devices.

Figure 4-23 shows the use of two XR-2240 timers in cascade to produce a very long duration timer. Unit 1 is used as the timebase, and has a frequency set by the expression $1/RC$. Only the eight-bit of the binary counter ($128T$) is used. The output of Unit 1 becomes the timebase of Unit 2, which has all of its outputs wired-OR together. The total period is $65\,536T$. When the RC time constant is 1 second, then the output duration is $65\,536$ seconds (18.2 hours) even though the stability and accuracy attributes are those of a 1 second timer.

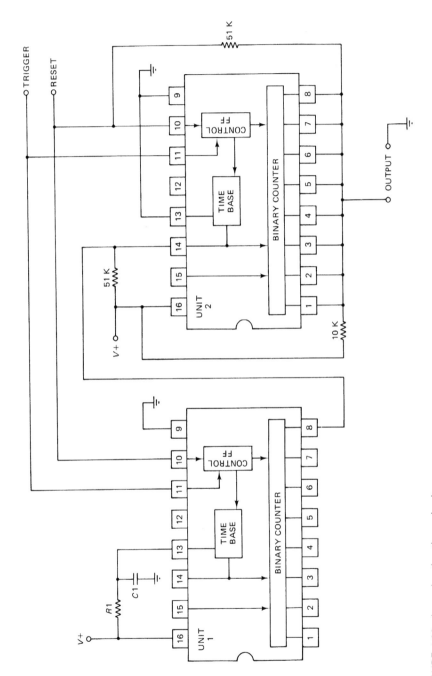

FIGURE 4-23 Long-duration timer circuit.

INTEGRATED CIRCUIT TIMERS

4-8 LM-122, LM-322, LM-2905 AND LM-3905 TIMERS

The LM-X22 and LM-X905 IC timers are precision devices that will operate with DC power supply voltages of +4.5 Vdc to +40 Vdc to produce durations of microseconds to hours. The LM-122 and LM-322 device package is shown in Fig. 4-24A, while that of the LM-2905 and 3905 is shown in Fig. 4-24B. The LM-2905/3905 devices are identical to the LM-122/322 except that the BOOST and V_{adj} pins are not available. The LM-122 and LM-2905 operate

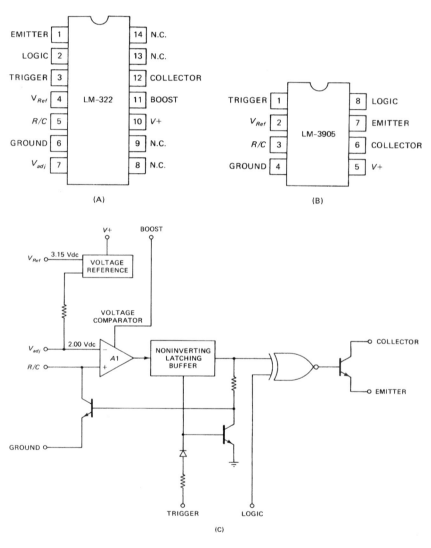

FIGURE 4-24 (A) LM-322 package and pinouts; (B) LM-3905 package and pinouts; (C) LM-322/LM-3905 timer internal circuit.

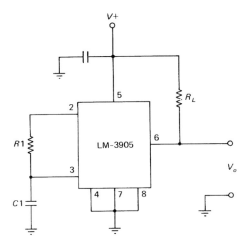

FIGURE 4-25 LM-3906 astable multivibrator.

over the temperature range $-55°C$ to $+125°C$, while the LM-322 and LM-3905 operates over $0°C$ to $+70°C$.

The internal circuitry of these timers is shown in Fig. 4-24C. The RC timing network is monitored by the noninverting input of a voltage comparator (A1); the inverting input is biased to +2.00 Vdc. The output of the comparator is routed to a noninverting latching buffer, which in turn drives an Exclusive-OR (XOR) gate. The alternate input of the XOR gate is connected to the LOGIC pin on the device. The output of these devices is a floating emitter and floating collector transistor. Both ground referenced and floating loads, at potentials up to 40 Vdc, can be accommodated with this arrangement. The V_{adj} pin can be used in the LM-122 and LM-322 devices to vary the timing ratio up to 50:1 by using an external voltage. The circuit for the basic timer is shown in Fig. 4-25. The output duration is set by $T = R1C1$.

4-9 SUMMARY

1. The 555 timer is RC timed and can be used as either a monostable multivibrator (MMV) or astable multivibrator (AMV). Most 555 devices are available in the 8-pin miniDIP package, although some wider temperature range devices are sold in the 8-pin metal can IC package.

2. The 555 MMV is triggered by bringing pin no. 2 to drop from above $2(V+)/3$ to less than $(V+)/3$. The trigger signal can be a regular pulse, or a switch that shorts a normally HIGH level to ground potential.

3. The 555 is normally nonretriggerable. It can be made retriggerable by adding an external discharge transistor to dump the charge across the timing capacitor in response to a trigger pulse.

INTEGRATED CIRCUIT TIMERS

4. The XR-2240 timer is similar to the 555 in the timebase section. The XR-2240 contains a binary counter that is driven by the timebase. The outputs of the binary counter are open-collector NPN transistors, so external pull-up resistors are needed. Times from $1RC$ to $255RC$ can be achieved by setting the timebase (RC) and connecting the required outputs together in a wired-OR configuration. Very long duration timers (e.g. $65\,536RC$) can be made by cascading XR-2240 devices.

4-10 RECAPITULATION

Now return to the objectives and Pre-quiz questions at the beginning of the chapter and see how well you can answer them. If you cannot answer certain questions, place a check mark to each and review the appropriate parts of the text. Next, try to answer the questions and work the problems below, using the same procedure.

4-11 STUDENT EXERCISES

1. Design, build and test a 555 monostable multivibrator (Fig. 4-5C) with an output duration of 10 milliseconds. Examine the output waveforms and the capacitor voltage waveforms together on a dual trace oscilloscope. Use various values of capacitance and resistance to see what happens to the duration.
2. Build the circuit of Exercise 1 above, using a +5 Vdc power supply. Use a squarewave as the trigger signal, making sure that it has adequate amplitude to cause triggering. Examine the output pulse and the trigger pulse on a dual trace oscilloscope, and note their relationship.
3. Design, build and test a retriggerable monostable multivibrator. Test the circuit using a squarewave input signal to the trigger pin.
4. Design, build and test an astable multivibrator using the 555 (Fig. 4-16B). Select component values for an oscillating frequency of approximately 1000 Hz. Initially make $R1 = R2$. Examine the capacitor voltage and output waveforms on a dual trace oscilloscope. Vary the values of $R1$ and $R2$ to create different duty cycles. Also vary $R1$, $R2$ and $C1$ and examine both oscillating frequency and duty factor.

4-12 QUESTIONS AND PROBLEMS

1. The 555 is an example of an _____ -timed multivibrator circuit.
2. Draw a block diagram of the 555 internal circuit and explain its operation.
3. Draw the circuit diagram for a 555 monostable multivibrator.
4. Draw the circuit diagram for a 555 astable multivibrator.
5. An RC circuit is made consisting of a 0.0015 µF capacitor, and a 12 kohms resistor. What is the RC time constant?
6. A 555 timer is used to produce an 8 ms output pulse. Calculate the resistance required if a 0.01 µF capacitor is used.

7. A series *RC* network consisting of a 0.01 μF capacitor and a 100 kohms resistor is connected to a +12 Vdc voltage source. What is the time required for the capacitor voltage to rise from +2.5 Vdc to +7.5 Vdc?

8. A timer is designed like the 555, but with the internal resistor voltage divider consisting of the following resistors: $R_a = 6.8$ kohms, $R_b = 5$ kohms, and $R_c = 10$ kohms. Calculate the potentials applied to the inputs of *COMP*1 and *COMP*2. Derive the timing equation for monostable operation.

9. A 555 timer is used to make an astable multivibrator with a frequency of 500 ms. Calculate the resistances needed if the duty cycle is 75% and the timing capacitor is 0.01 μF.

10. An XR-2240 timer is used to make a ten-minute timer. What outputs must be wired-OR together if the timebase uses the following timing components: $R1 = 1\,000\,000$ and $C1 = 10$ μF?

11. Why is the 555 class IC timer free of operating frequency drift when the $V+$ DC power supply voltage slowly changes?

12. Define the rule for operating the *trigger* input of the 555 in the monostable multivibrator mode.

13. The output pin of the 555 timer is capable of sinking or sourcing _____ mA of current.

14. Define the operation of the *reset* pin on the 555.

15. Draw a circuit that will allow retriggerable operation of the 555 in the monostable multivibrator mode.

16. Draw the circuit and timing diagram for a missing pulse detector based on the 555.

17. Draw the block diagram for a simple tachometer circuit based on the 555 that will produce a DC output that is proportional to the frequency of an input pulse signal.

18. Draw the block diagram for a pulse stretcher circuit that will increase a 100 μs pulse to 10 ms.

19. Draw the circuit for a variable duty cycle astable multivibrator based on the 555.

20. Draw the circuit for a 555 sawtooth generator.

21. Draw the circuit for an XR-2240 monostable multivibrator that has a duration of 255 seconds.

22. All eight outputs of the XR-2240 are connected together with a 10 kohm pull-up resistor. Calculate the output duration of this monostable multivibrator when the *RC* time constant of the timing network is 0.5 ms.

CHAPTER 5

IC data converter circuits and their application

OBJECTIVES

1. Understand the basic elements of data conversion.
2. Be able to describe the operation of the standard $R-2R$ DAC.
3. Be able to describe the operation of the principal forms of A/D converter.
4. Be able to design circuits using common IC data converter devices.

5-1 PRE-QUIZ

These questions test your prior knowledge of the material in this chapter. Try answering them before you read the chapter. Look for the answers (especially those you answered incorrectly) as you read the text. After you have finished studying the chapter try answering these questions again, and those at the end of the chapter (see Section 5-9).

1. A _____ DAC has an external reference voltage source.
2. A DAC-08 device has a reference potential of +7.5 Vdc, and an reference input resistance of 10 kohms. Calculate the output current at full scale (binary input = 11111111_2).
3. Describe how a DAC can be used to digitally synthesize a sawtooth waveform.
4. What is the output current of a DAC-08 when the reference current is 2 mA and the applied binary word is 10000000_2?

   (NOTE: Binary (base-2) numbers are used in digital circuits. The notation adopted in this text is to add a subscript 2 to indicate a binary number. Thus,

1000 would mean one-thousand, while 1000_2 means a binary number that evaluates to decimal eight.)

5-2 BASICS OF DATA CONVERSION

The data converter does one of two jobs, it either: (1) converts a binary digital word to an equivalent analog current or voltage, or, (2) converts an analog current or voltage to an equivalent binary word. The former are called digital-to-analog converters (DACs), while the latter are called analog-to-digital converters (A/D or ADC). These devices form the interface between digital computers and the analog world. This chapter will examine the basic functioning of these data conversion building blocks.

5-3 APPROACHES TO DATA CONVERSION

There are several approaches to converting data from electronic circuits and scientific instruments into digital binary numbers required by a digital computer. This job requires an A/D converter. The opposite type of converter (DAC) does the opposite work. It converts the binary output from a computer into an analog voltage or current. In discussing data converters it is best to begin with a discussion of DACs because many A/D circuits contain embedded DACs.

5-4 DIGITAL-TO-ANALOG CONVERTERS (DACS)

There are several different approaches to DAC design, but all of them are varieties of a weighted current or voltage system that accepts binary words by appropriate switch contacts. The most common example is the *R–2R ladder* shown in Fig. 5-1. The active element, $A1$, is an operational amplifier in a unity gain inverting follower configuration. In the circuit of Fig. 5-1 the digital inputs are shown as mechanical switches, but in a real data converter circuit the switches would be replaced by electronic switching devices (e.g. transistors). The electronic switches are driven by either a binary counter, or an N-bit parallel data line.

A precision reference voltage source (V_{ref}) is required for accurate data conversion. This voltage is most often +2.56 volts, +5.00 volts or +10.00 volts. Other voltages can also be used, however. The accuracy of the converter is dependent upon the precision of the reference voltage source. There are other sources of error, but if the reference voltage accuracy is poor, then there is no hope for any other factors to be effective in improving the performance of the circuit. Although almost any voltage regulator can be pressed into service as the reference, it is prudent to select a precision, low-drift model.

Returning to Fig. 5-1, consider the circuit action under circumstances where various binary bits are either HIGH or LOW. If all bits are LOW,

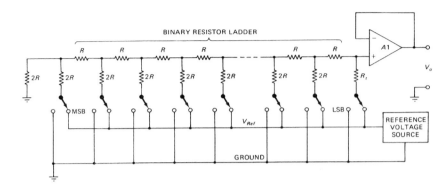

FIGURE 5-1 R–2R resistor network DAC circuit.

then the output voltage will be zero. The value of the output voltage is given by the product $I \times R$, and when all bits are LOW this current is zero. In practical circuits, though, there might be some output voltage under these circumstances due to offsets in the operational amplifier, the R–$2R$ ladder and the electronic switches. These offsets can be nulled to zero output voltage when all bits are intentionally set to zero (or ignored, if negligible).

The unterminated R–$2R$ ladder produces an output current. Some commercial IC DACs are current output models, and have no output amplifier. If there is a terminating resistor (R_t) shunting the output terminals of the DAC, then the circuit produces an output voltage $I_o \times R_t$. The output impedance of such a circuit tends to be high, so some of these DACs use an output amplifier in order to produce a low-impedance voltage output. The transfer function of the R–$2R$ ladder type of DAC is:

$$V_o = \frac{V_{ref} A}{2^N} \tag{5-1}$$

where:
    V_o is the output potential
    V_{ref} is the reference potential
    A is the decimal value of the applied binary word
    N is the number of bits in the applied binary word

EXAMPLE 5-1

An eight-bit R–$2R$ DAC has a 10 volt reference potential. What is the output voltage when the applied binary word is 11000000_2 (decimal 192)?

Solution

$$V_o = \frac{V_{ref} A}{2^N}$$

$$V_o = \frac{(10 \text{ Vdc})(192)}{2^8}$$

$$V_o = \frac{1920}{256} = 7.5 \text{ Vdc} \qquad \blacksquare$$

If the most significant bit (MSB) is made 1 (i.e. HIGH), then the output voltage will be approximately $V_{\text{ref}}/2$. Similarly, if the next most significant bit is turned on (set to HIGH) and all others are LOW, then the output will be $V_{\text{ref}}/4$. The least significant bit (LSB) would contribute $V_{\text{ref}}/2^N$ to the total output voltage. For example, with an 8-bit DAC, the LSB changes the output $V_{\text{ref}}/2^8$, or $V_{\text{ref}}/256$. This change is called the *1-LSB value* because it is the change that occurs in response to a change in the least significant bit. Figure 5-2 graphs the output of a voltage DAC in response to the entire range of binary numbers applied to the digital inputs. The result is a staircase waveform that rises by the 1-LSB value for each 1-LSB change of the binary word. This step height represents the minimum discernible resolution of the circuit.

The reference source can be either internal or external to an integrated circuit DAC. If the reference voltage or current source is external, then the DAC is said to be a *multiplying DAC* or MDAC. The multiplication takes place between the analog reference and the fraction defined as $A/2^N$ in Eq. (5-1). If the reference source is completely internal, and not adjustable (except for fine trimming), then the DAC is said to be a *non-multiplying DAC* or simply DAC.

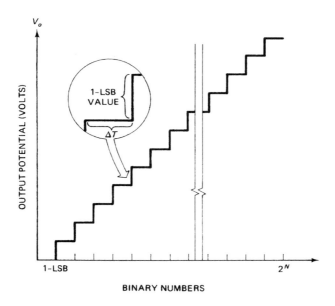

FIGURE 5-2   DAC output function.

5-4.1 Coding schemes

There are several different coding methods for defining the transfer function of the DAC. The most common of these are: *unipolar positive, unipolar negative, symmetrical bipolar* and *asymmetrical bipolar*.

The unipolar coding schemes provide an output voltage of one polarity only. These circuits usually produce 0 volts for the minimum and some positive or negative value for the maximum. Because one binary number state represents 0 Vdc, there are $[2^N - 1]$ states to represent the analog voltages within the range. For example, in an 8-bit system there are 256 states, so if one state (00000000_2) represents 0 Vdc then there are 255 possible states for the non-zero voltages. Thus, the maximum output voltage is always 1-LSB less than the reference voltage. For example, in a 10.00 Vdc system, the maximum output voltage is 10.00 volts/256 = 0.039 volts = 39 mV less. Thus, the maximum output voltage will be approximately 9.96 Vdc. The unipolar positive coding scheme is:

Unipolar output voltage	Binary input word
0.00 Vdc	00000000_2
$V_{max}/2$	10000000_2
V_{max}	11111111_2

The negative version of this coding scheme (unipolar negative) is identical, except that the midscale voltage is $-V_{max}/2$ and the fullscale output voltage is $-V_{max}$. A variant on the theme inverts the definition so that:

Unipolar output voltage	Binary input word
0.00 Vdc	11111111_2
V_{max}	10000000_2
V_{max}	$00000000_{@-\{2\}}$

The bipolar coding scheme faces a difficulty that requires a trade-off in the design. There is an even number of output states in a binary system. For example, in the standard 8-bit system there are 256 different output states. If one state is selected to represent 0 Vdc, then there are 255 states left to represent the voltage range. As a result, there is an even number of states to represent positive and negative states either side of 0 Vdc. For example, 127 states might be assigned to represent negative voltages, and 128 to represent positive voltages. In the asymmetrical bipolar coding, therefore, the pattern might look like:

Bipolar output voltage	Binary input word
$-V_{max}$	00000000_2
0.00 Vdc	10000000_2
$+V_{max}$	11111111_2

A decision must be made regarding which polarity will lose a small amount of dynamic range.

The other bipolar coding system is the symmetrical bipolar scheme. The decision in the symmetrical scheme is that each polarity will be represented by the same number of binary states either side of 0 Vdc. But this scheme does not permit a dedicated state for zero. The scheme is:

Bipolar output voltage	Binary input word
$-V_{max}$	00000000_2
$-$Zero (-1 LSB)	01111111_2
0.00 Vdc	(disallowed)
$+$Zero ($+1$ LSB)	10000000_2
$+V_{max}$	11111111_2

The state 'plus zero' is more positive by 1-LSB value than 0 Vdc, while the 'minus zero' state is more negative than 0 Vdc by the same 1-LSB value.

5-4.2 A practical IC DAC example

A number of different manufacturers offer low-cost IC DACs that contain almost all of the circuitry needed for the process, except possibly the reference source (although some devices do contain the reference source also) and some operational amplifiers for either level shifting or current to voltage conversion.

For purposes of this chapter the DAC-08 device is used as a practical circuit example. This eight-bit DAC is now something of an industry standard, and is available from several sources. This DAC is sometimes designated LMDAC-0800. An easily available, and closely related, device is the DAC-0806.

Figure 5-3A shows the basic circuit configuration for the DAC-08. In subsequent circuits, shown later in this chapter, the power supply terminals are deleted for simplicity's sake; they will always be the same as shown here. The internal circuitry of the DAC-08 is the $R-2R$ ladder shown in the previous section, but has two outputs: I_o and NOT-I_o. These current outputs are unipolar and complementary (Fig. 4-3B); if the full-scale output current is I_{max}, then $I_{max} = [I_o + \text{NOT-}I_o]$. The specified value of I_{max} on the DAC-08 family of DACs is 2 mA.

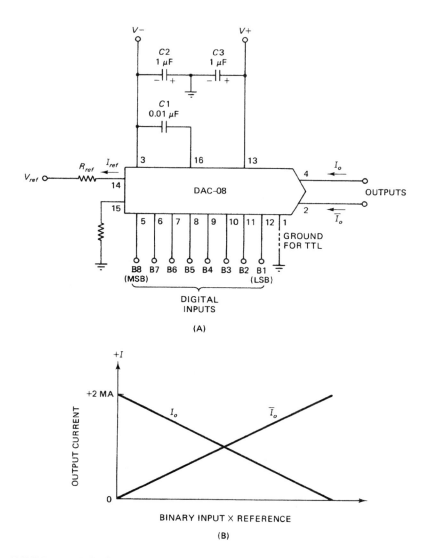

FIGURE 5-3 (A) DAC-08 device circuit; (B) output functions.

Two types of input signal are required to make this DAC work: an *analog reference* and an *8-bit digital signal*. The analog signal is the reference current, I_{ref}, applied through pin no. 14. This current may be generated by combining a precision reference voltage source with a precision, low temperature coefficient, resistor to convert V_{ref} to I_{ref}. Alternatively, a constant current source may be used to provide I_{ref}. For TTL compatibility of the binary inputs, make $V_{\text{ref}} = 10.000$ volts, and $R_{\text{ref}} = 5000$ ohms.

The other type of input is the eight-bit digital word, which is applied to the IC at pins 5 through 12, as shown. The logic levels which operate these

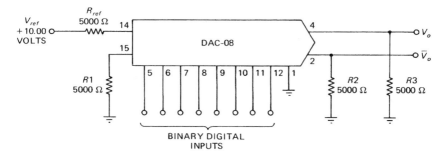

FIGURE 5-4   DAC-08 circuit.

inputs can be preset by the voltage applied to pin no. 1 (for TTL operation, pin no. 1 is grounded). In the TTL-compatible configuration shown, LOW is 0 to 0.8 volts, while HIGH is +2.4 to +5 volts.

Figure 5-4 shows the connection of the DAC-08 (less power supply and reference input) required to provide the simplest form of unipolar operation over the range of approximately 0 to 10 volts. When the input word is 00000000_2, then the DAC output is 0 volts, plus-or-minus the DC offset error. A half-scale voltage (−5 volts) is given when the input word is 10000000_2. This situation occurs when the MSB is HIGH and all other digital inputs are LOW. The full-scale output will exist only when the input word is 11111111_2 (all HIGH). The output under full-scale conditions will be −9.96 volts, rather than 10 volts as might be expected (note: 9.96 volts is 1-LSB less than 10 volts).

The circuit in Fig. 5-4 works by using resistors $R2$ and $R3$ as current-to-voltage converters. When currents I_o and NOT-I_o pass through these resistors, a voltage drop of IR, or $5.00 \times I_o$ (mA) is created. A problem with this circuit is that is has a high source impedance (5 kohms, with the values shown for $R2/R3$).

Figure 5-5 shows a simple method for converting I_o to an output voltage (V_o) with a low output impedance (less than 100 ohms) by using an inverting follower operational amplifier. The output voltage is simply the product of the output current and the negative feedback resistor:

$$V_o = R \times I_o \qquad (5\text{-}2)$$

As in the case previously described, a 5000 ohm resistor and a 10.00 Vdc reference voltage will produce a 9.96 volt output voltage when the DAC-08 is set up for TTL inputs and 2.0 mA $I_{o(\max)}$.

The frequency response of the DAC circuit can be tailored to meet certain requirements. The normal output waveform of the DAC is a staircase when the digital input increments up from 00000000_2 to 11111111_2 in a monotonic manner. In order to make the staircase into an actual ramp function, a low-pass filter is needed at the output to remove the 'stepness' of the normal

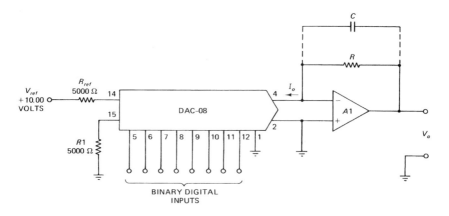

FIGURE 5-5 Low-impedance output DAC-08 circuit using inverting follower amplifier.

waveform. A capacitor shunted across the feedback resistor (R) will offer limited filtering on the order of −6 dB/octave above a cutoff frequency of:

$$F = \frac{1\,000\,000}{2\pi RC_{\mu F}} \tag{5-3}$$

where:
F is the −3 dB frequency in hertz (Hz)
R is in ohms (Ω)
$C_{\mu F}$ is in microfarads (μF)

In most practical circuits the required value of F is known from the application. It is the highest frequency Fourier component in the input waveform. It is necessary to calculate the value of capacitor needed to achieve that cut off frequency, so would swap the F and C terms:

$$C_{\mu F} = \frac{1\,000\,000}{2\pi RF} \tag{5-4}$$

A related method shown in Fig. 5-6 produces an output voltage of the opposite polarity from that of Fig. 5-5. The circuit of Fig. 5-6 is connected to a noninverting unity gain follower at the output. The output voltage is the product of I_o and R2. If a higher output voltage is needed, then the circuit variant shown in the inset to Fig. 5-6 can be used. In this case, the output amplifier has gain, so the output voltage will be:

$$V_o = I_o \times R2 \times \left[\frac{R4}{R3} + 1\right] \tag{5-5}$$

One of the ways to achieve bipolar binary operation is shown in Fig. 5-7. In this circuit the output amplifier is a DC differential amplifier, and both

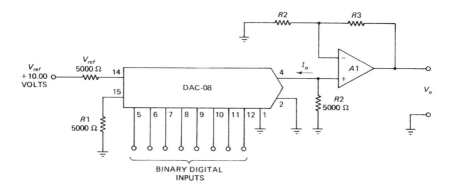

FIGURE 5-6 DAC-08 low impedance output circuit using noninverting follower amplifier.

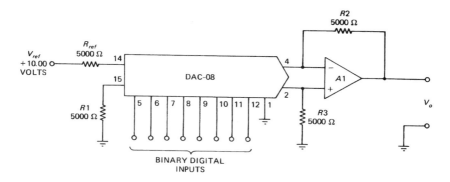

FIGURE 5-7 DAC-08 low impedance output circuit using differential amplifier.

current outputs of the DAC-08 are used. Note that the maximum and minimum voltages are positive and negative. The zero selected can be either (+)zero (i.e. +1-LSB voltage), or (−)zero (i.e. −1-LSB voltage). It cannot be exactly zero because an even number of output codes are equally spaced around zero. In other words, the absolute value of FS(−) is equal to the absolute value of FS(+). There are also circuits that make zero = zero, but at the expense of uneven ranges for FS(−) and FS(+).

A practical DAC circuit is shown in Fig. 5-8. This circuit combines the circuit fragments shown earlier to make a complete circuit that can be used. The power connections are not shown. The heart of this circuit is a DAC-08 connected in the bipolar binary circuit discussed above.

The reference potential in Fig. 5-8 is a REF-01 10.000 volt IC reference voltage source. Potentiometer $R1$ adjusts the value of the actual voltage, and also serves as a full-scale adjustment for the output voltage, V_o.

The output amplifier can be a 741-class operational amplifier, or any other form; the need is not critical. Potentiometer $R9$ acts as a zero adjustment for

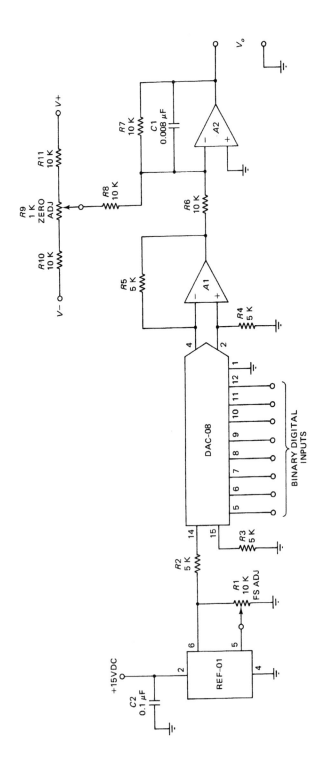

FIGURE 5-8 Practical DAC-08 circuit.

158 IC DATA CONVERTER CIRCUITS AND THEIR APPLICATION

V_o. The capacitor across $R7$ limits the frequency response to 200 Hz with the value shown. This frequency limit can be changed with the equation given earlier.

Adjustment
1. Set the binary inputs all LOW (00000000_2).
2. Adjust $R9$ for $V_o = 0.00$ volts.
3. Set all binary inputs HIGH (11111111_2).
4. Adjust potentiometer $R1$ for $V_o = 9.96$ volts.

5-4.3 Digital synthesis sawtooth generator

In this chapter digitally synthesized versions of waveform generators are discussed, with special emphasis on sawtooth signal generators.

Sawtooth signal generators are used for a variety of purposes in electronics: electronic music synthesizers, sweeping RF oscillators, audio signal generators, and voltage controlled oscillators (VCO), certain laboratory bench tests, calibrating oscilloscopes, and other applications. In addition, there are many circuit applications for embedded sawtooth generators. Examples include situations where the sawtooth is used to provide precision calibration of an oscilloscope timebase. If the sawtooth used to sweep the oscilloscope horizontally is controlled from a stable crystal oscillator, then a very precise sweep rate is possible. The 'standard' solid-state sawtooth generator circuit consists of an operational amplifier Miller integrator circuit excited by a squarewave.

The circuit for a digitally synthesized sawtooth generator is shown in Fig. 5-9. The heart of this circuit is $IC1$, a DAC-0806 eight-bit DAC. This DAC is related to the DAC-08.

A DAC-0806 produces an output current that is proportional to: (a) the reference voltage or current, and (b) the binary word applied to the digital inputs. The controlling function for the DAC is:

$$I_o = I_{ref} \times \frac{A}{256} \tag{5-6}$$

where:
  I_o is the output current from pin no. 4
  I_{ref} is the reference current applied to pin no. 14
  A is the decimal value of the binary word applied to the eight binary inputs (pins 5–12)

The reference current is found from Ohm's law, and is the quotient of the reference voltage and the series resistor at pin no. 14. In analog data acquisition and display systems, the reference voltage is a precision, regulated potential. But in this case the precision may not be needed, so use the $V+$ power supply is used as the reference voltage. Therefore, the reference

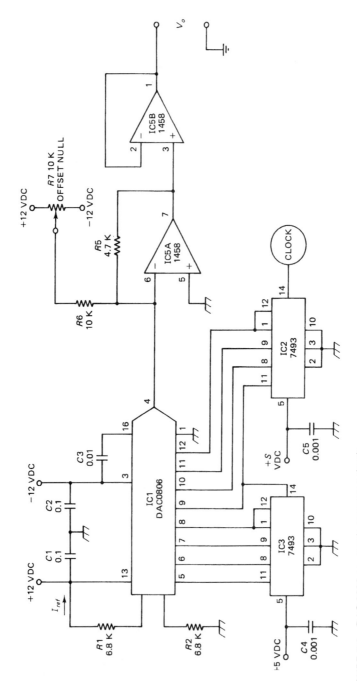

FIGURE 5-9 DAC-0806 sawtooth generator circuit.

current is $(+12 \text{ VDC})/R1$. With the value of $R1$ shown (6800 ohms), I_{ref} is 0.00018 amperes, or 1.8 mA. Values from 500 μA to 2 mA are permissible with this device.

The reference current sets the maximum value of output current, I_o. When a full-scale binary word (11111111_2) is applied to the binary inputs, the output current I_o is:

$$I_o = 1.8 \text{ mA} \times \frac{255}{256} \tag{5-7}$$

$$I_o = 1.8 \text{ mA} \times 0.996 = 1.78 \text{ mA}$$

The DAC-0806 is a current output DAC, so an op-amp current-to-voltage converter is required in order to make a sawtooth voltage output function. Such a circuit is an ordinary inverting follower without an input resistor. The output voltage (V_o) will rise to a value of ($I_o \times R5$).

The waveform produced by this circuit is shown in Fig. 5-10. This waveform has a period of about 5 ms (200 Hz), and an amplitude of about 3 volts.

The actual output waveform is 'staircased' in binary steps equal to the 1-LSB current of the DAC, or the 1-LSB voltage, is the smallest step change in output potential cause by changing the least significant bit (B1) either from 0 to 1, or 1 to 0. The reason why the steps in Fig. 5-10 cannot be seen is that the frequency response of the 741 operational amplifier used for the current-to-voltage converter acts as a low-pass filter to smooth the waveform. If a higher frequency op-amp is used, then a capacitor shunting $R3$ will serve to low-pass filter the waveform. A -3 dB frequency (F) of 1 or 2 kHz will suffice to smooth the waveform. The value of the capacitor is calculated from:

$$C_{\mu F} = \frac{1\,000\,000}{2\pi RF} \tag{5-8}$$

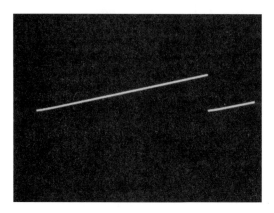

FIGURE 5-10 Output waveform of sawtooth generator.

where:

$C_{\mu F}$ is the capacitance in microfarads (μF)
F is the -3 dB cutoff frequency in hertz (Hz)
$R3$ is expressed in ohms (Ω)

Select a clock frequency that is 256 times the desired sawtooth fundamental frequency.

Other waveforms. The simple binary counter that allows the sawtooth to be generated also limits the circuit. It can be modified, however, to produce any other waveform. For a triangle waveform replace the straight binary counters with a base-16 up/down binary counter. Arrange the digital control logic to reverse the direction of the count when the maximum state (11111111_2) is sensed.

There are two ways to generate waveforms other than a sawtooth or triangle, and both of them involve using a computer memory. The binary bit pattern representing the waveform is stored in memory, and then output in the right sequence. One method uses a *read only memory* (ROM) chip which is pre-programed with the bit pattern representing the waveform (Fig. 5-11). A binary counter circuit connected as an address generator selects the bit pattern sequence.

The second method is to store the bit pattern in a computer, and then output it under program control via an eight-bit parallel output port. This method is usable for both generating special waveforms, and for linearizing the tuning characteristic of circuits such as VCOs, swept oscillators, etc. The digital solution to the linearization problem (Chapter 3) involves storing a look-up correction table in either a ROM or computer memory.

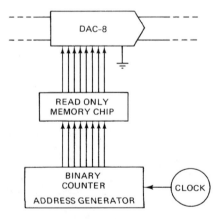

FIGURE 5-11 Any waveform can be generated by storing its binary pattern in a read only memory (ROM) chip.

5-5 ANALOG-TO-DIGITAL CONVERTERS (A/D)

The analog-to-digital converter (A/D) is used to convert an analog voltage or current input to an output binary word that can be used by a computer. Of the many techniques that have been published for performing an A/D conversion, only a few are of interest to us; so we will consider only the *voltage-to-frequency, single-slope integrator, dual-slope integrator, counter* (or *servo*), *successive approximation* and *flash* methods.

5-5.1 Integration A/D methods

Most digital panel meters (DPM) and digital multimeters (DMM) use either the *single integration* or *dual-slope integration* methods for the A/D conversion process. An example of a single-slope integrator A/D converter is shown in Fig. 5-12A. The single-slope integrator is simple, but is limited to those applications that can tolerate accuracy of 1-2%.

The single-slope integrator A/D converter of Fig. 5-12A consists of five basic sections: *ramp generator, comparator, logic, clock* and *output encoder*. The ramp generator is an ordinary operational amplifier Miller integrator with its input connected to a stable, fixed, reference voltage source. This makes the input current I_{ref} essentially constant; so the voltage at point B will rise in a nearly linear manner, creating the voltage ramp.

The comparator is an operational amplifier that has no feedback loop. The circuit gain is the open-loop gain (A_{vol}) of the device selected — typically very high even in low-cost operational amplifiers. When the analog input voltage V_x is greater than the ramp voltage, the output of the comparator is saturated at a logic-HIGH level.

The logic section consists of a main AND gate, a main-gate generator, and a clock. The waveforms associated with these circuits are shown in Fig. 5-12B. When the output of the main-gate generator is LOW, switch $S1$ remains closed, so the ramp voltage is zero. The main-gate signal at point A is a low frequency squarewave with a frequency equal to the desired time-sampling rate. When point A is HIGH, $S1$ is open, so the ramp will begin to rise linearly. When the ramp voltage is equal to the unknown input voltage V_x, the differential voltage seen by the comparator is zero; so its output drops LOW.

The AND gate requires all three inputs to be HIGH before its output can be HIGH also, from times $T0$ to $T1$, the output of the AND gate will go HIGH every time the clock signal is also HIGH.

The encoder, in this case an eight-bit binary counter, will then see a pulse train with a length proportional to the amplitude of the analog input voltage. If the A/D converter is designed correctly, then the maximum count of the encoder will be proportional to the maximum range (full-scale) value of V_x.

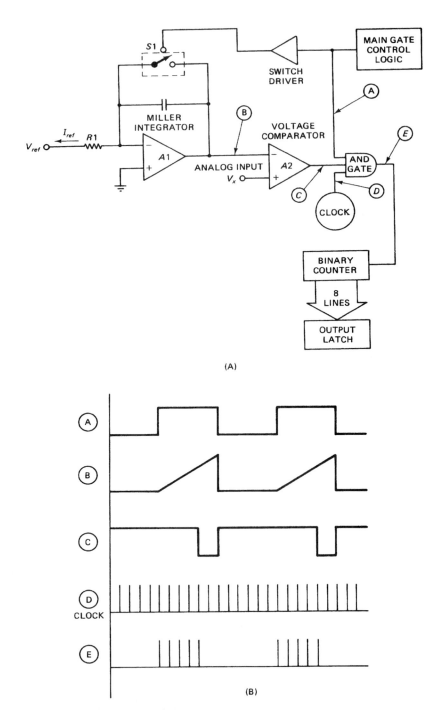

FIGURE 5-12 (A) Single-slope integrator circuit; (B) timing waveforms.

Several problems are found in single-slope integrator A/D converters:

- The ramp voltage may be nonlinear.
- The ramp voltage may have too steep or too shallow a slope.
- The clock pulse frequency could be wrong.
- It may be prone to changes in apparent value of V_x caused by noise.

Many of these problems are corrected by the dual-slope integrator of Fig. 5-13. This circuit also consists of five basic sections: *integrator, comparator, control logic section, binary counter*, and *a reference current* or *voltage source*. An integrator is made with an operational amplifier connected with a capacitor in the negative feedback loop, as was the case in the single-slope version. The comparator in this circuit is also the same sort of circuit as was used in the previous example. In this case, though, the comparator is ground-referenced by connecting +IN to ground.

When a *start* command is received, the control circuit resets the counter to 00000000_2, resets the integrator to 0 volts (by discharging $C1$ through switch $S1$), and sets electronic switch $S2$ to the analog input (position A). The analog voltage creates an input current to the integrator which causes the integrator output to begin charging capacitor $C1$; the output voltage of the integrator will begin to rise. As soon as this voltage rises a few millivolts above ground potential (0 Vdc) the comparator output snaps HIGH-positive. A HIGH comparator output causes the control circuit to enable the counter, which begins to count pulses.

The counter is allowed to overflow, and this output bit sets switch $S2$ to the reference source (position B). The graph of Fig. 5-13B shows the integrator charging during the interval between start and the overflow of the binary counter. At time $T2$ the switch changes the integrator input from the analog signal to a precision reference source. Meanwhile, at time $T2$ the counter had overflowed, and again it has an output of 00000000_2 (maximum counter +1 more count is the same as the initial condition). It will, however, continue to increment so long as we have a HIGH comparator output. The charge accumulated on capacitor $C1$ during the first time interval is proportional to the average value of the analog signal that existed between $T1$ and $T2$.

Capacitor $C1$ is discharged during the next time interval ($T2-T3$). When $C1$ is fully discharged the comparator will see a ground condition at its active input, so will change state and make its output LOW. Even though this causes the control logic to stop the binary counter, it does not reset the binary counter. The binary word at the counter output at the instant it is stopped is proportional to the average value of the analog waveform over the interval ($T1-T2$). An *end-of-conversion* (EOC) signal is generated to notify the computer so that it knows the output data is both stable and valid (therefore ready for use).

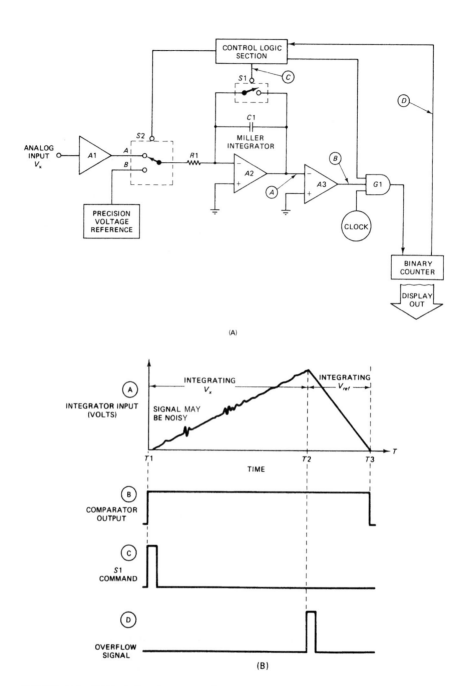

FIGURE 5-13 (A) Dual-slope integrator; (B) timing waveforms.

5-5.2 Voltage-to-frequency converters

These circuits are not A/D converters in the strictest sense, but are very good for representing analog data in a form that can be tape recorded on a low-cost audio machine, or transmitted over radio. The V/F converter output can also be used for direct input to a computer if a binary counter is used to measure the output frequency. Two forms of V/F converter are common. One is a voltage controlled oscillator (VCO); that is, a regular oscillator circuit in

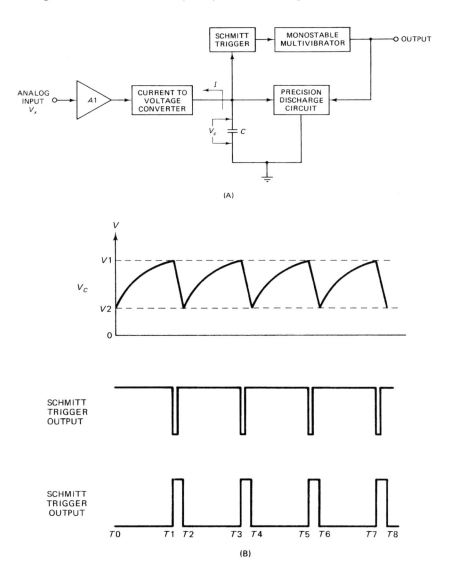

FIGURE 5-14 V-to-F converter: (A) block diagram; (B) timing.

which the output frequency is a function of an input control voltage. If the VCO is connected to a binary or binary coded decimal (BCD) counter, then the VCO becomes a V/F form of A/D converter.

The type of V/F converter shown in Fig. 5-14 is superior to the VCO method. The circuit is shown in Fig. 5-14A, while the timing waveforms are shown in Fig. 5-14B. The operation of this circuit is dependent upon the charging of a capacitor, although not an RC network as in the case of some other oscillator or timer circuits. The input voltage signal (V_x) is amplified (if necessary) by $A1$, and then converted to a proportional current level in a voltage-to-current converter stage. If the voltage applied to the input remains constant, so will the current output of the V-to-I converter (I).

The current from the V-to-I converter is used to charge the timing capacitor, C. The voltage appearing across this capacitor (V_c) varies with time as the capacitor charges (see the V_c waveform in Fig. 5-14B). The precision discharge circuit is designed to discharge capacitor C to a certain level ($V2$) whenever the voltage across the capacitor reaches a predetermined value ($V1$). When the voltage across the capacitor reaches $V2$, a Schmitt trigger circuit is fired that turns on the precision discharge circuit. The precision discharge circuit, in its turn, will cause the capacitor to discharge rapidly but in a controlled manner to value $V1$. The output pulse snaps HIGH when the Schmitt trigger fires (i.e. the instant V_c reaches $V1$) and drops LOW again when the value of V_c has discharged to $V2$. The result is a train of output pulses whose repetition rate is exactly dependent upon the capacitor charging current, which, in turn, is dependent upon the applied voltage. Hence, the circuit is a voltage-to-frequency converter.

Like the VCO circuit, the output of the V/F converter can be applied to the input of a binary counter. The parallel binary outputs become the data lines to the computer. Alternatively, if the frequency is relatively low the computer can be programmed to measure the period between pulses. Also, certain computer interface chips have built-in timers that can measure the period.

5-5.3 Counter type (servo) A/D converters

A counter type A/D converter (also called 'servo' or 'ramp' A/D converters) is shown in Fig. 5-15. It consists of a *comparator, voltage output DAC, binary counter*, and the necessary *control logic*. When the start command is received, the control logic resets the binary counter to 00000000_2, enables the clock, and begins counting. The counter outputs control the DAC inputs; so the DAC output voltage will begin to rise when the counter begins to increment. As long as analog input voltage V_{in} is less than V_{ref} (the DAC output), the comparator output is HIGH. When V_{in} and V_{ref} are equal, however, the comparator output goes LOW, which turns off the clock and stops the counter. The digital word appearing on the counter output at this time represents the value of V_{in}.

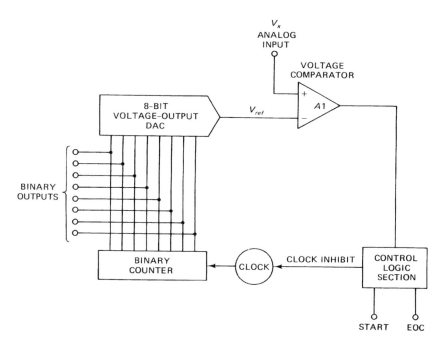

FIGURE 5-15 Binary counter A/D converter circuit.

Both slope and counter type A/D converters take too long for many applications, on the order of 2^N clock cycles (where N = number of bits). Conversion time becomes critical if a high frequency component of the input waveform is to be faithfully reproduced. Nyquist's criteria requires that the sampling rate (i.e. conversions per second) be at least twice the highest Fourier frequency to be recognized.

5-5.4 Successive approximation A/D converters

Successive approximation A/D conversion is best suited for many applications where speed is important. This type of A/D converter requires only $N + 1$ clock cycles to make the conversion, and some designs allow truncation of the conversion process after fewer cycles if the final value is found prior to $N + 1$ cycles.

The successive approximation converter operates by making several successive trials at comparing the analog input voltage with a reference generated by a DAC. An example is shown in Fig. 5-16. This circuit consists of a *comparator, control logic section*, a *digital shift register, output latches* and a *voltage output DAC*.

When a START command is received, a binary 1 (HIGH) is loaded into the MSB of the shift register, and this sets the output of the MSB latch

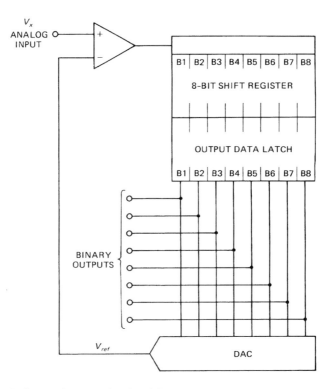

FIGURE 5-16 Successive approximation A/D converter.

HIGH. A HIGH in the MSB of a DAC will set the output voltage V_{ref} to half-scale. If the input voltage V_{in} is greater than V_{ref}, the comparator output stays HIGH and the HIGH in the shift register MSB position shifts one-bit to the right and therefore occupies the next most significant bit (bit 2). Again the comparator compares V_{in} with V_{ref}. If the reference voltage from the DAC is still less than the analog input voltage, the process will be repeated with successively less significant bits until either a voltage is found that is equal to V_{in} (in which case the comparator output drops LOW) or the shift register overflows.

If, on the other hand, the first trial with the MSB indicates that V_{in} is less than the half-scale value of V_{ref}, the circuit continues making trials below V_{ref}. The MSB latch is reset to LOW and the HIGH in the MSB shift register position shifts one-bit to the right to the next most significant bit (bit 2). Here the trial is repeated again. This process will continue as before until either the correct level is found, or overflow occurs. At the end of the last trial (bit 8 in this case) the shift register overflows and the overflow bit becomes an end-of-conversion (EOC) flag to tell the rest of the world that the conversion is completed.

This type, and most other types of A/D converters, requires a starting pulse and signals completion with an EOC pulse. This requires the computer or other digital instrument to engage in bookkeeping to repeatedly send the start command and look for the EOC pulse. If the start input is tied to the EOC output, then conversion is continuous, and the computer need only look for the periodic raising of the EOC flag to know when a new conversion is ready. Such operation is said to be *asynchronous*.

5-5.5 Parallel or `flash´ A/D converters

The parallel A/D converter (Fig. 5-17) is probably the fastest A/D circuit known. Indeed, the very fastest ordinary commercial products use this

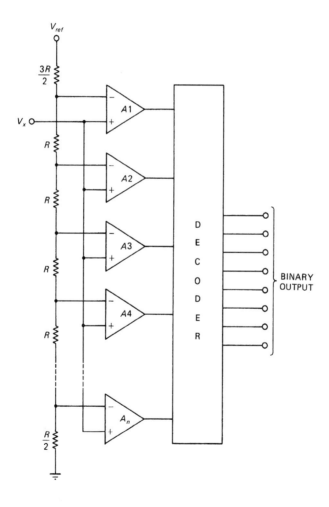

FIGURE 5-17   Flash A/D converter.

method. Some sources call the parallel A/D converter the 'flash' circuit because of its inherent high speed.

The parallel A/D converter consists of a bank of $[2^N - 1]$ voltage comparators biased by reference potential V_{ref} through a resistor network that keeps the individual comparators 1-LSB apart. Since the input voltage is applied to all the comparators simultaneously, the speed of conversion is limited essentially by the slew rate of the slowest comparator in the bank, and also by the decoder circuit propagation time. The decoder converts the output code to binary code needed by the computers.

5-6 SUMMARY

1. An analog-to-digital (A/D) converter produces a binary number output that represents the applied analog input voltage or current. A digital-to-analog converter (DAC) is exactly the opposite: it outputs an analog voltage or current that is proportional to an applied binary number.
2. Most DAC designs are based on the $R-2R$ binary weighted ladder. If the output is unterminated, then the DAC is a current output type; if the output is terminated in either a load resistor or an operational amplifier, then the DAC is a voltage output type.
3. Several different coding schemes are used in data converters: *unipolar positive*, *unipolar negative*, *symmetrical bipolar* and *asymmetrical bipolar* are the most common.
4. IC DACs can be used to generate waveforms. If the DAC digital inputs are incremented by an ordinary binary counter, then a sawtooth is generated; if incremented by an up/down counter, then a triangle waveform is generated; if the binary inputs are driven from a read only memory (ROM), that is in turn driven by a binary counter used as an address generator, then any waveform whose binary bit pattern can be stored in ROM can be created.
5. Several different approaches are used for A/D conversion: *voltage-to-frequency converter*, *single-slope integrator*, *dual-slope integrator*, *counter* (or *servo*), *successive approximation* and *flash* methods.

5-7 RECAPITULATION

Now return to the objectives and Pre-quiz questions at the beginning of the chapter and see how well you can answer them. If you cannot answer certain questions, place a check mark to each and review the appropriate parts of the text. Next, try to answer the questions and work the problems below, using the same procedure.

5-8 STUDENT EXERCISES

1. Build a 5-bit $R-2R$ ladder such as Fig. 5-1 from resistors in the 100 to 1000 ohms range (NOTE: $R_t \gg R$). Test the circuit for the output voltage when various binary numbers are applied by closing the associated switches.

2. Build a working DAC for 0 to +9.96 Vdc operation using a DAC-08, MC-1408, DAC-0806 or similar commercial IC DAC. Test the circuit for the output voltage when various binary numbers are applied to the eight binary input lines.
3. Configure the circuit in the previous exercise as an AC programmable gain amplifier.
4. Build an 8-bit A/D converter based on the DAC circuit of exercise no. 2 above.

5-9 QUESTIONS AND PROBLEMS

1. Draw the circuit for an $R-2R$ ladder DAC with voltage output.
2. A DAC-08 device has a reference potential of +10 Vdc, and an reference input resistance of 5 kohms. Calculate the output current at fullscale (binary input = 10000000_2).
3. Describe how a DAC can be used to digitally synthesize a triangle waveform. Use a circuit diagram to illustrate the case.
4. Describe a method, and provide a simplified circuit diagram, for a synthesizer based on a DAC that will output the human electrocardiogram waveform.
5. What is the output current of a DAC-08 when the reference current is 2 mA and the applied binary word is 00000001_2?
6. What is the full-scale output of a DAC-08 IC digital-to-analog converter when the reference potential is +10.00 volts DC?
7. An eight-bit DAC is designed for symmetrical bipolar operation. If the full-scale outputs are ±5.00 Vdc, what are the values of +ZERO and −ZERO?
8. In addition to the binary digital inputs the _____ DAC has an external reference voltage.
9. The circuit of Fig. 5-4 uses a DAC-08. What is the output voltage V_o if I_o = 1.25 mA?
10. In the circuit of Fig. 5-5, what is the output voltage if $R = 4.7$ kohms and $I_o = 2$ mA?
11. List the different forms of A/D converter circuit.
12. Draw a block diagram for a servo converter circuit.
13. A _____ A/D converter has a conversion time of 2^N clock cycles.
14. A _____ A/D converter has a conversion time of $[2^N - 1]$ clock cycles.
15. A _____ A/D converter has a conversion time of $[N + 1]$ clock cycles.
16. The _____ A/D converter is generally regarded as the fastest form of converter, even though the output must be decoded before it can be used in binary systems.
17. Draw the output waveform of a 3-bit DAC. Label the 1-LSB value.
18. Draw the output circuit for a DAC-08 that will produce a low output impedance and a maximum potential of +10 Vdc when $I_o = 0.002$ amperes.
19. What is the purpose of capacitor C in Fig. 5-5?
20. In Fig. 5-7, what is the output current of the DAC-08 when $V_o = +10$ Vdc?

21. What is the purpose of the REF-01 IC in Fig. 5-8?
22. Draw the circuit for a 3-bit flash A/D converter.
23. An 8-bit A/D converter is designed to convert a range of potentials from 0 to +4.5 volts. (a) What is the 1-LSB value? (b) What voltage is represented by 11000010_2?
24. Discuss the terms 'plus zero' and 'minus zero' with respect to a bipolar coded A/D converter.
25. Write the binary codes for zero, mid-scale and full-scale output potentials from a unipolar coded DAC.

CHAPTER 6

Audio applications of linear IC devices

OBJECTIVES

1. Be able to list various audio applications for linear IC devices.
2. Understand the methods of design used for audio circuits.
3. Learn the types of linear ICs typically used in audio circuits.

6-1 PRE-QUIZ

These questions test your prior knowledge of the material in this chapter. Try answering them before you read the chapter. Look for the answers (especially those you answered incorrectly) as you read the text. After you have finished studying the chapter try answering these questions again, and those at the end of the chapter (see Section 6-11).

1. Draw the circuit for a 'bridge' audio power amplifier.
2. An audio preamplifier for a two-way radio transmitter must have a gain of 100, and a frequency response of 300 Hz to 3300 Hz between -3 dB points. Calculate appropriate values for the input and feedback resistors, and the value of the capacitor shunting the feedback resistor. Assume a noninverting follower.
3. In the circuit above, the input circuit contains a transformer with a 1:2 turns ratio. What is the total voltage gain of this circuit?
4. Draw the circuit for a three-channel audio mixer with an inverting characteristic.

6-2 AUDIO CIRCUITS

The audio frequency range is generally accepted to be the nominal range of human hearing: 20 Hz to 20 000 Hz. Although some applications require frequency ranges lower than 20 Hz, or higher than 20 kHz, most audio circuits fall within the stated range. Some applications use a smaller range, however. For example, communications equipment sometimes uses 300 Hz to 3000 Hz as the frequency range. Similarly, an AM radio station is allowed to transmit audio frequencies up to 5000 Hz, but not higher. Thus, for audio equipment designed for use in AM radio stations the audio frequency range may be limited to 20 Hz to 5000 Hz. Cassette tape recorders are often limited to 16 kHz at the upper end, while other forms of tape recorder may use the entire 20 kHz range.

There are also some applications where a nominal 20 kHz frequency range actually requires a higher range of frequencies. High-fidelity enthusiasts once debated the desirability of 100 kHz or even 250 kHz as the upper end of the frequency range. The argument is based on the fact that all non-sinusoidal waveforms are composed of a fundamental sinewave frequency plus a number of harmonic sine and cosine waves added to it. The tonal coloration of any given musical instrument is dependent on which 'overtones' (i.e. harmonics) are present, and what amplitude they exhibit. That is why a violin playing an 'A' note at 440 Hz sounds different from a piano or trumpet playing the same 440 Hz note. The difference between these instruments is the harmonic content of the waveform. It was argued by some that the 'brightness' or 'coloration' of the audio is believed to be compromised if the harmonics above 20 kHz are eliminated or attenuated.

There are several different classes of amplifier used in audio circuits. The *preamplifier* is designed to accept a low-level voltage signal and boost it to a higher voltage level. For example, the output of a phonograph cartridge is on the order of 1 to 5 mV, while the standard 'line out' level required to drive a power amplifier is on the order of 100 mV to 500 mV. Some tape recorder magnetic heads produce outputs in the 100 μV to 500 μV range. The preamplifier is designed to boost the signal from these levels to a level from ×10 to ×1000 higher. A *control amplifier* is designed to operate at low gain (×1 to ×10), but performs certain signals processing functions such as bass and treble boost, balance, or some other similar function. Finally, a *power amplifier* is designed to boost a 100 mV to 500 mV voltage signal to a power level sufficient to drive a loudspeaker or other load. Audio power amplifiers tend to produce output levels from 100 milliwatts to hundreds of watts. In a typical audio system these circuits are arranged as in Fig. 6-1.

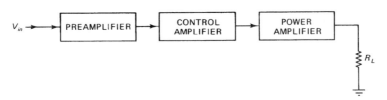

FIGURE 6-1 Block diagram of audio power amplifier.

6-3 PREAMPLIFIER CIRCUITS

A preamplifier circuit is an audio amplifier that gives some initial amplification to the signal before passing it to another circuit for additional amplification or other processing. For example, a microphone has a low-level output (a few millivolts for a dynamic type, and up to 0.2 volts for a crystal type). A preamplifier typically boosts the microphone signal to 100 to 1000 millivolts before it is applied to the input of a power amplifier (for the loudspeaker) or the input of a transmitter modulator (depending upon use).

Microphone preamplifier. Figure 6-2 shows a simple gain-of-100 microphone preamplifier for communications and non-Hi-Fi public address uses. This circuit uses an LM-301 operational amplifier in the noninverting follower gain configuration. With the values of feedback and 'input' resistors ($R1$ and $R2$) shown, the gain is 101.

The input circuit of this preamplifier is capacitor-coupled to the microphone. In order to keep the input bias currents of the op-amp from charging the capacitor (and thereby latching up the op-amp), a 2.2 megohm resistor ($R3$) is connected from the noninverting input to ground. This circuit is relatively general, but can be modified towards the less complex if we use a

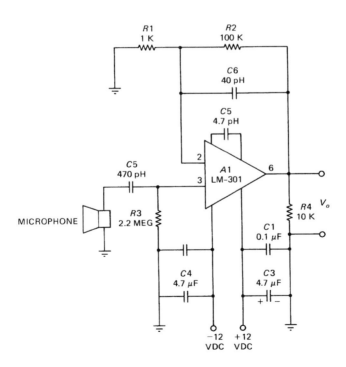

FIGURE 6-2 Microphone preamplifier.

AUDIO APPLICATIONS OF LINEAR IC DEVICES

dynamic microphone only. Those forms of microphone use either a high or low impedance coil (like a loudspeaker, but reversed in function) that is permanently connected into the circuit. In that case, delete $R3$ and $C7$, and connect the microphone between ground and pin no. 3 of the op-amp. If the microphone is to be disconnected from time to time, however, keep $R3$ in the circuit to prevent the op-amp output from saturating at or near $V+$ when the noninverting input 'goes open'.

The frequency response of this circuit is tailored by $C5$ and also by the capacitor shunting the feedback resistor ($C6$). With the values of capacitance shown, the upper -3 dB point in the response curve is a little over 3000 Hz, and falls off at a rate of approximately -6 dB/octave above that frequency. The equation for determining the frequency response is:

$$f = \frac{1}{2\pi RC} \qquad (6\text{-}1)$$

where:
  f is the -3 dB frequency of the amplifier in hertz (Hz)
  R is the resistance in ohms (Ω)
  C is the capacitance in farads (F)

EXAMPLE 6-1

Calculate the capacitance (in μF) needed to establish a -3 dB frequency response of 5000 Hz when the feedback resistor is 100 kohms.

Solution

$$C6 = \frac{1}{(2)(3.14)(100\,000\ \Omega)(5000\ \text{Hz})} \times \frac{10^6\ \mu F}{1\ F}$$

$$C6 = \frac{1}{3.14 \times 10^9} \times \frac{10^6}{1\ F} = 3.2 \times 10^{-4}\ \mu F \qquad \blacksquare$$

Frequency tailored preamplifiers are found in several different varieties that are characterized by the shape of the frequency response. Figure 6-3 shows three common response characteristics. The form shown in Fig. 6-3A is the single-slope version. The gain (A_v) is essentially flat from DC (or some low AC frequency) to some higher frequency, F_c. Above this frequency the gain rolls off at a specified rate such as -6 dB/octave. The frequency response is determined as the point at which the gain drops off -3 dB from the low frequency response. The -3 dB point may be the natural roll-off frequency of the amplifier being used, or it may be artificially tailored as in the example above.

A two-slope response characteristic is shown in Fig. 6-3B. In this case, there are two gain segments. A higher gain (A_{V1}) is found from the lower end of the response range to frequency $F1$, while a lower gain (A_{V2}) is found from $F1$ to F_c. This type of response is commonly found in tape recorder

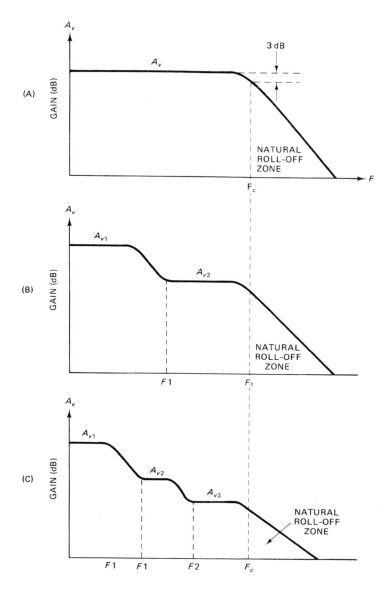

FIGURE 6-3 Frequency tailored preamplifier responses: (A) normal; (B) dual-slope; (C) triple-slope.

systems. The reason it is used is to roll-off the high frequency preemphasis imparted to the tape recorded audio in order to overcome noise. If the gain of the playback preamplifier wasn't tailored in a manner such as Fig. 6-3B, then the reproduced audio would be unbalanced towards the treble end of the range.

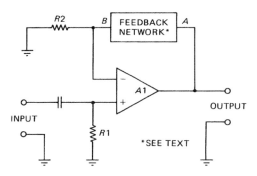

FIGURE 6-4 Typical 'generic' preamplifier circuit.

A three-slope version is shown in Fig. 6-3C. In this case there are three different gains: A_{V1}, A_{V2} and A_{V3}. This characteristic is used to restore the audio balance of phonograph records when reproduced.

A typical preamplifier circuit is shown in Fig. 6-4. The basic circuit is a noninverting follower operational amplifier circuit. The input network $R1/C1$ sets the lower -3 dB frequency according to Eq. (6-1). Resistor $R1$ is needed if there is a substantial bias current from the input of the operational amplifier. That current could charge the capacitor ($C1$), causing a DC offset that will eventually saturate the output of the amplifier. Resistor $R1$ drains the DC charge that would otherwise accumulate on $C1$. Some modern operational amplifiers have such a low value of input bias current (picoamperes) that $R1$ can sometimes be omitted. The input impedance of the preamplifier is set by resistor $R1$, so it should have a value as high as possible. The general rule of thumb is to make the input resistor not less than ten times the source impedance of the transducer or other device used to originate the audio signal. Standard audio amplifiers use a 600 ohm output impedance (a holdover from telephone technology), and in those situations almost any value of input resistor is sufficient. However, tape heads tend to have a source impedance on the order of 20 kohms to 100 kohms, while crystal microphones and phonograph cartridges are over 100 kohms. Thus, it is common practice to use a value of 1 megohms to 10 megohms for $R1$.

The characteristic of the upper -3 dB roll-off is set by the feedback network shown in Fig. 6-4 as a block. Figure 6-5 shows several different forms of RC feedback network circuit. If a single resistor is used, then the roll-off will be single-slope (Fig. 6-3A), but the -3 dB point is the natural -3 dB point for the amplifier. If a network such as Fig. 6-5B is used, then the characteristic is still single-slope, but the -3 dB point is set by Eq. (6-1). A two-slope characteristic (Fig. 6-3B) is created by the network of Fig. 6-5C. In this case, the low frequency gain setting resistor ($R4$) is shunted by a roll-off network that has a -3 dB 'breakpoint' determined by Eq. (6-1). Finally, the three-slope network is shown in Fig. 6-5D.

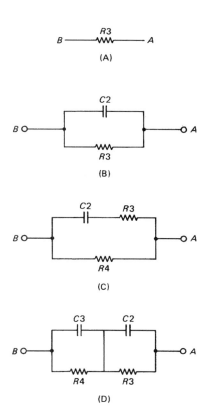

FIGURE 6-5 Feedback networks for preamplifier: (A) normal uncompensated; (B) single-slope; (C) two-slope; (D) three-slope.

6-3.1 LM-381 IC preamplifier

There are a number of operational and non-operational amplifier audio preamplifiers on the market. One of the earliest was the MC-1303P device, which used two gain-of-20 000 wideband operational amplifiers in a single package. Two devices are required for stereo audio applications. Another class of non-operational amplifiers are represented by the LM-381 device shown in Fig. 6-6. This device consists of a pair of differential devices similar to operational amplifiers, but operating from a single DC power supply of +9 to +40 Vdc. The LM-381 offers voltage gains of 112 dB per channel with a unity gain bandwidth product of 15 MHz. The power bandwidth is 75 kHz at an output signal of 20 volts peak-to-peak. The device boasts a total harmonic distortion (THD) of 0.1% (1 kHz, $A_v = 60$ dB).

The LM-381 offers a power supply rejection ratio (PSRR) of 120 dB, and a stage to stage isolation (at 1000 Hz) of 60 dB. The LM-381 is internally frequency compensated, yet can be custom tailored through external

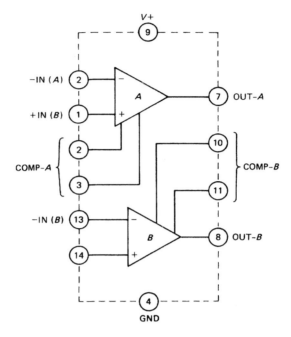

FIGURE 6-6 LM-1303 two-channel audio preamplifier.

compensation as well. The device also features output short circuit protection. The output terminals will sink 2 mA, or source 8 mA, of current. The total equivalent noise input is 0.5 to 1 V_{rms}.

The basic circuit for the LM-381 device is shown in Fig. 6-7A. The input signal is applied directly to the noninverting input through a DC blocking capacitor, while the gain and frequency response characteristic is set by the feedback network between output and inverting input. Examples of feedback networks is shown in Figs 6-7B (NAB tape preamplifier) and 6-7C (RIAA phonograph preamplifier).

An inverting amplifier configuration of the LM-381 is shown in Fig. 6-8. In this circuit the noninverting input is decoupled to ground through a capacitor ($C2$), while the input signal is applied to the inverting input through an RC network ($C1R1$). The gain is set by a feedback network consisting of a resistor voltage divider ($R2/R3$) and a series resistor ($R4$).

A low-distortion version of this same circuit is shown in Fig. 6-9. The THD of the more general circuit (above) is on the order of 0.1%, while that of Fig. 6-9 is 0.05% or less for a gain (A_v) of ten. In this circuit the gain setting circuit is the feedback network $R2/R3$ connected to the inverting input. This circuit is basically the same as the inverting operational amplifier configuration. The input signal is applied, however, not through $R3$ but rather through an RC network ($R1C1$) similar to the previous circuit.

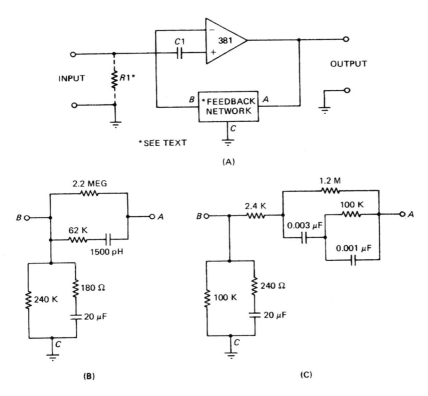

FIGURE 6-7 (A) LM-381 preamplifier circuit; (B) NAB tape preamplifier; (C) RIAA phono preamplifier.

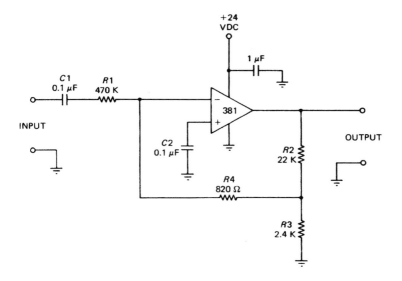

FIGURE 6-8 LM-381 inverting amplifier configuration.

AUDIO APPLICATIONS OF LINEAR IC DEVICES **183**

CMOS preamplifier. Figure 6-10 shows two general purpose preamplifiers based on the CA-3600E device. This IC is not an operational amplifier, but rather it is a complimentary COS/MOS transistor array. A single stage design is shown in Fig. 6-10A, while a multi-stage design is in Fig. 6-10B. The internal transistor array equivalent circuit for one transistor pair is shown

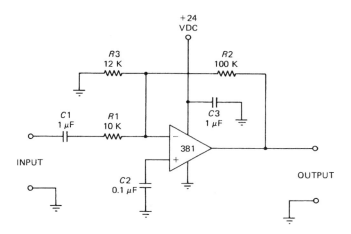

FIGURE 6-9 Low-distortion LM-381 circuit.

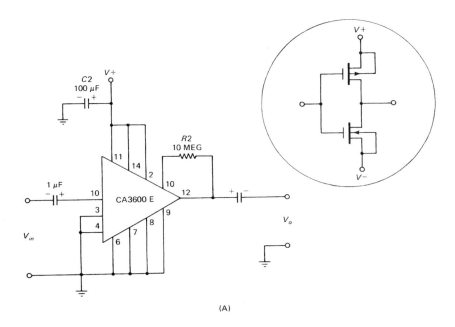

(A)

FIGURE 6-10   (A) Single-stage wideband preamplifier using CA-3600E device; (B) multistage amplifier.

184   AUDIO APPLICATIONS OF LINEAR IC DEVICES

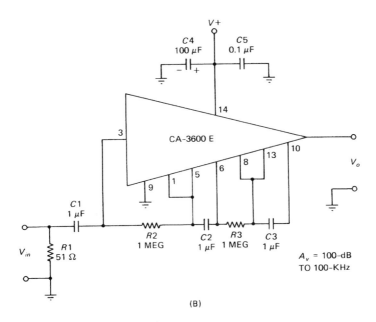

(B)

FIGURE 6-10 (continued)

in the inset to Fig. 6-10A. The single-stage design is capable of up to 30 dB of gain at a V+ of 15 volts DC, and slightly more at lower potentials (but only at a sacrifice of the 1 MHz −3 dB point).

The multi-stage design shown in Fig. 6-10B is capable of voltage gains to 100 dB at frequencies up to 1 MHz (assuming a 10 volt DC supply — gain drops to 80 dB for +15 volt supplies). This gain and frequency response is quite useful in audio and other applications, but must be approached with caution when actually built. Be sure to keep the input and output sides of the amplifier separated as much as possible. Also make sure that the power supply decoupling capacitors are mounted as close as possible to the body of the IC.

The 50 ohm input impedance of this amplifier takes it out of the audio amplifier category (which usually use higher system impedances), unless a preamplifier stage is provided that will have a higher impedance. A good candidate is the circuit of Fig. 6-10A.

6-3.2 600-Ohm audio circuits

Professional audio and broadcast applications generally use a 600-ohm balanced line between devices in the system. For example, a remote preamplifier will have a 600-ohm balanced output and will connect to the next stage through three-wire line. Such a system uses two 'hot' signal lines and a ground line as the stage-to-stage interconnection. An amplifier with a

600-ohm balanced output is usually called a *line driver amplifier*, while an amplifier with a 600-ohm balanced input is a *line receiver amplifier*. Some amplifiers are both line drivers and line receivers.

Figure 6-11A shows a line receiver amplifier based on the LM-301 operational amplifier used in the unity gain, noninverting configuration. The input circuit is a line transformer. If the turns ratio of transformer $T1$ is 1:1, then the overall gain of the circuit is unity. But if the transformer has a turns ratio other than 1:1, then the 'gain' is essentially the turns ratio of the transformer. For example, suppose a transformer is selected with a 600-ohm balanced input winding, and a 10 000 ohm secondary winding. Such a transformer will have a sec/pri impedance ratio of 10 000/600, or about 17:1, so the turns

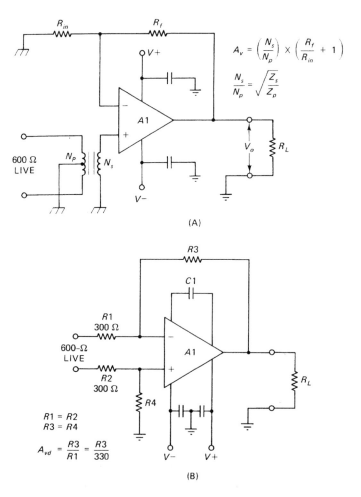

FIGURE 6-11 (A) Transformer coupled 600-ohm balanced line receiver amplifier; (B) DC-coupled 600-ohm line receiver amplifier; (C) 600-ohm line driver amplifier.

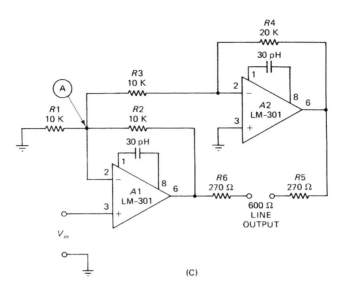

(C)

FIGURE 6-11 (continued)

ratio is on the order of the squareroot of 17, or about 4.1. Thus, the voltage actually applied to the noninverting input will be 4.1 times higher than the input signal voltage from the 600-ohm line.

$$A_\text{V} = \left[\frac{N_\text{s}}{N_\text{p}}\right]\left[\frac{R_\text{f}}{R_\text{in}} + 1\right] \quad (6\text{-}2)$$

Like most other op-amp circuits, Fig. 6-11A requires two DC power supplies ($V-$ and $V+$). Typically, these power supplies can be anywhere from ±6 to ±15 volts. As in other circuits, each DC power supply is decoupled with a pair of capacitors. The output of this circuit is an ordinary single-ended voltage amplifier (as in other op-amp circuits), so will typically have a very low impedance. Some designers use a transformer coupled output circuit. It is possible to get away with a 600-ohm 1:1 transformer if the natural output impedance of the op-amp is on the order of 50 ohms or so. The general rule of thumb is that the primary impedance of any transformer selected should be ≥10 × the natural output impedance of the device for best voltage transformation.

Another way to make a 600-ohm line input amplifier is to use the simple DC differential circuit, and make sure the input resistors are 300 ohms each (see Fig. 6-11B). The gain of the circuit is R3/300, provided that $R1 = R2 = 300$ ohms, and $R3 = R4$.

Figure 6-11C shows a line driver amplifier based on a pair of operational amplifiers (the DC power supply connections are deleted for simplicity). The output circuitry is balanced because it is made from two single-ended op-amps driven out of phase with each other. The low output impedance of the operational amplifier, plus the 270 ohm series resistance, makes the balanced output impedance a total of approximately 600 ohms.

The circuit of Fig. 6-11C is a good example of the clever use of one of the properties of the ideal op-amp. Recall that one of the ideal properties is that inputs 'stick together'. In other words, applying a voltage to one input causes the same voltage to appear at the other input. In this case, for example, the AF input signal voltage applied to the noninverting input of amplifier A1 also appears on the inverting input of that same amplifier. Thus, V_{in} appears on both the noninverting input and at point A in Fig. 6-11C. Therefore, the circuit uses point A to feed the other half of the balanced circuit, amplifier A2. Because A1 is a noninverting gain-of-two circuit, and A2 is a gain-of-two inverting circuit, the two sides are out of phase with each other — which is the condition required of the two 'balanced' output lines.

6-4 AUDIO MIXERS

An *audio mixer* is a circuit that linearly combines audio signals from two or more inputs into a single channel. Application examples include multiple microphone public address systems, multiple guitar systems (music) or radio station audio consoles where inputs from tape players, record players, and two or more microphones are combined into a single line that goes to the transmitter's modulator input.

Audio mixer I. One form of audio mixer is an operational amplifier version shown in Fig. 6-12. This circuit is basically nothing more than a unity gain inverting follower with multiple inputs. Three audio lines are identified here: V1, V2 and V3. Each of these sources are applied to the input of the operational amplifier, and see gains of R4/R1, R4/R2 and R4/R3, respectively. Because all resistors are 100 kohms, the gains for all three channels are unity.

Gain can be customized on a channel by channel basis by varying the input resistance value. The gain of any given channel will be 100 kohm/R, where R is the input resistance (R1, R2 or R3) in kohms. The output voltage for multiple inputs mixers is given by:

$$V_o = R4 \times \left[\frac{V1}{R1} + \frac{V2}{R2} + \frac{V3}{R3} + \cdots + \frac{V_n}{R_n} \right] \qquad (6\text{-}3)$$

Be careful not to reduce the input resistance so much that the source is loaded. If the source is another operational amplifier preamplifier (or other voltage amplifier) then the input resistance can be reduced to several kohms without a problem. But if the source is a high impedance device, then 50 kohms is probably the minimum acceptable value.

In some cases, it is beneficial to increase the value of the feedback resistance to 1 megohm or so, in order to make the corresponding input resistances higher for any given gain. Remember, the input impedance seen by any single channel is the value of the input resistor.

A *master gain control* is provided by making feedback resistor R4 variable. If no gain control is needed, then make this resistor fixed. An audio taper

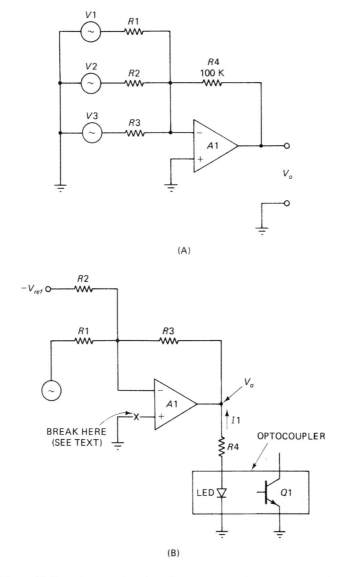

FIGURE 6-12 (A) Three-input audio mixer; (B) audio driver for optocoupler; (C) waveform with no DC offset; (D) waveform with DC offset.

AUDIO APPLICATIONS OF LINEAR IC DEVICES **189**

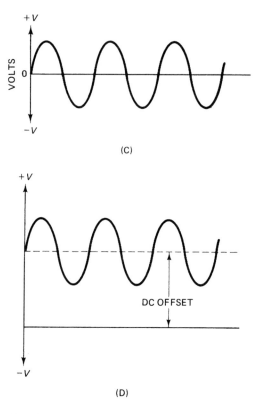

FIGURE 6-12 (continued)

potentiometer is used for most audio applications. If the application calls for a one-time 'set and forget' gain control adjustment (as might be true in radio station applications), then make R4 a trimmer potentiometer, otherwise it should be a panel type with a shaft appropriate for a knob (quarter inch half or full-round).

The operational amplifier selected can be almost any good op-amp with a gain bandwidth product sufficient for audio applications. Because the gain is unity, any GBW over 20 kHz will suffice (which means almost all devices except the 741 family — which will work in communications applications).

Audio mixer II. The circuit in Fig. 6-13 is an improved audio mixer based on the CA-3048 CDA amplifier array. This circuit provides approximately 20 dB of gain at each channel. The CA-3048 device is a 16-pin DIP integrated circuit that contains four independent AC amplifiers. Offering a gain of 53 dB with a GBW of 300 kHz (typical), the CA-3048 has a 90 kohm input impedance, and an output impedance less than 1 kohm. It will produce a maximum low-distortion output signal of 2 volts RMS, and can accept input signals up to 0.5 volts RMS.

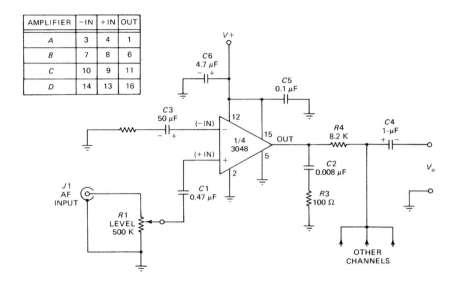

FIGURE 6-13 CA-3048 preamplifier circuit (up to four channels on single chip).

Each DC power supply can be up to +16 volts DC. There are two $V+$ and two ground connections. These multiple connections are used to reduce the internal coupling between amplifiers. The two $V+$ terminals are tied together externally, and the two ground terminals are tied together also. The $V+$ terminals are bypassed with a dual capacitor. $C5$ in Fig. 6-13 is a 0.1 µF unit, and is used for high frequencies, while $C6$ is a 4.7 µF tantalum electrolytic for low frequencies. Both capacitors must be mounted as close as possible to the CA-3048 body, with $C5$ taking precedence for closeness over $C6$ (high frequencies are more critical).

An RC network from the amplifier output to ground ($R3/C2$) is used to stabilize the amplifier, thus preventing oscillation. Like the power supply bypass/decoupling capacitors, these components need to be mounted as close as possible to the body of the amplifier.

Only one channel is shown in detail in Fig. 6-13 because of space limitations. But each of the other three channels are identical, and are joined with the circuitry shown at the output capacitor ($C4$) as shown. Each of the four channels has its own level control ($R1$), which also provides a high input impedance for the mixer.

6-5 COMPRESSION AMPLIFIER

A compression amplifier is one that reduces its gain on input signal peaks, and increases the gain in the signal valleys. These circuits are used by electronic music fans, and by broadcasters, to raise the average power in the signal without appreciable increase in total harmonic distortion. Figure 6-14 shows a compression amplifier circuit.

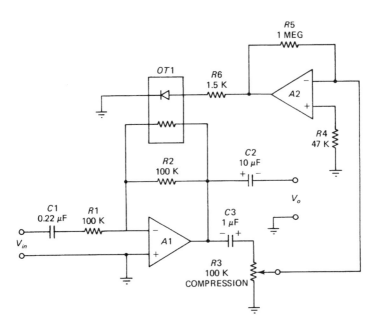

FIGURE 6-14 Compression amplifier circuit.

The amplifier ($A1$) is an audio operational amplifier, such as the LM-301 (power supply and compensation not shown). The gain of the circuit is set by the input resistor ($R1$), and a feedback resistance that consists of the parallel combination of $R2$ and the resistance of the optocoupler ($OT1$) output element. The $OT1$ resistance is set by the intensity of the light emitting diode (LED) brightness, which is, in turn, set by the signal amplitude produced by $A2$. Because the $A2$ output signal is proportional to the $A1$ output signal, overall gain 'reduces itself' (i.e. compresses). Optoisolator $OT1$ can be any resistance output device such as the Clairex, or a modern type (e.g. $H11$) that uses a JFET for the resistance element.

6-6 CONTROL AMPLIFIERS

Control amplifiers are, generally speaking, a category of circuits that will perform some processing function other than simple gain. Control amplifiers are generally low gain circuits ($\times 1$ to $\times 10$). It is common to find the output signal at the same amplitude as the input signal. In commercial and consumer audio equipment the primary functions of control amplifiers are *tone control, input source selection* and *channel balance* (in stereo equipment).

6-6.1 Tone control circuits

The tone control circuit either accentuates (boosts) or attenuates a particular band of frequencies within the audio spectrum. The simplest tone control circuit is the *treble roll-off* circuit of Fig. 6-15A. This method is used only in the cheapest equipment. The basis for the treble roll-off control is an *RC* network shunted across the audio line. The *RC* time constant of the active section of $R1$ (a) and $C1$: $T = R1(a) \times C1$. The apparent 'tone change' is dependent upon this time constant, and sets the -3 dB frequency and the roll-off slope (see Fig. 6-15B).

The treble roll-off control functions only by attenuating the high frequencies. Thus, the 'bass' setting does not boost the low frequencies, but rather it gives the illusion of bass by reducing the amplitude of signals in the treble range. Similarly, the 'treble' setting of the control merely restores the original tone balance, not boost the high frequency ranges.

A *shelf equalizer* will boost or cut a specific range of frequencies in either the bass or treble ranges. This type of circuit produces a characteristic such

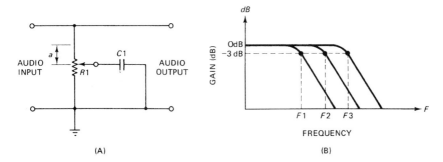

FIGURE 6-15 Treble roll-off tone control: (A) circuit (B) frequency responses.

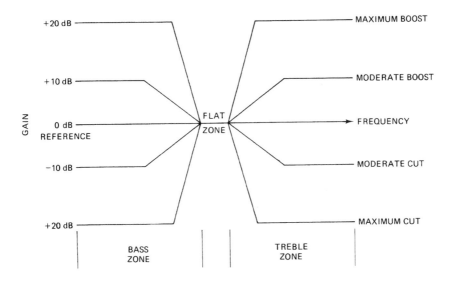

FIGURE 6-16 Shelf equalizer frequency response.

AUDIO APPLICATIONS OF LINEAR IC DEVICES **193**

as Fig. 6-16. The 0 dB line represents the *flat* condition in which the signal is neither boosted nor cut. In the *boost* condition, frequencies within the defined range (e.g. treble or bass) are amplified more than frequencies within the flat zone. In the cut condition the signals within the band are amplified less than those in the flat zone.

In simple equipment only two bands are affected by the tone controls: *bass* (low) and *treble* (high). In *graphic equalizers* there are more controls, each of which can either boost or cut. Each control operates over a narrower range of audio frequencies than simple bass and treble controls.

A circuit for a single band in a shelf equalizer is shown in Fig. 6-17A. It consists of a unity gain inverting follower based on an operational amplifier.

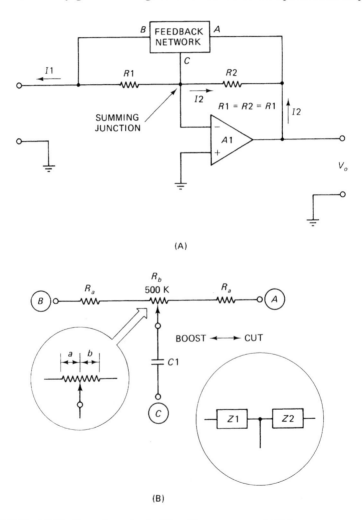

FIGURE 6-17 (A) Shelf equalizer circuit; (B) treble range control; (C) bass range control.

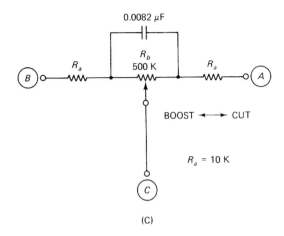

FIGURE 6-17 (continued)

In this circuit $R1 = R2 = R$. The flat zone gain is $-R2/R1 = -R/R = -1$. This gain figure is modified at certain frequency ranges by a frequency sensitive feedback network that shunts $R2/R1$.

A treble range feedback network is shown in Fig. 6-17B. The bass version of this circuit is shown in Fig. 6-17C. This circuit is similar, except that a capacitive reactance shunts the control. Let's evaluate the treble version. The network consists of two equal resistances (R_a) in series with a potentiometer (R_b). The total resistance of the network is $2R_a + R_b$. The wiper of the potentiometer is connected to the summing junction of the inverting follower through a capacitor, $C1$. The reactance of $C1$ is $X_{C1} = 1/2\pi FC1$, so one may conclude that the frequency selectivity is a function of a circuit resistances and X_{C1}.

The potentiometer (R_b) serves as the front panel adjustment control. The resistance of the 'pot' is divided into two portions (labelled 'a' and 'b'). The relative magnitudes of R_{ba} and R_{bb} is determined by the setting of the pot's adjustment shaft. When the shaft is set to the midpoint, $R_{ba} = R_{bb}$.

Input current $I1$ is determined by input signal voltage V_{in} and the parallel combination of $R1$ and the impedance ($Z1$) consisting of R_a, R_{b1} and X_{C1}. The feedback current $I2$ is determined by output voltage V_o and the parallel combination of resistance $R2$ and the impedance ($Z2$) consisting of R_a, R_{bb} and X_{C1}.

Inverting amplifiers such as Fig. 6-17A can be evaluated by appealing to the fact that the summing junction is at ground potential by virtue of the fact that the noninverting input is grounded.

If R_b is set to the mid-point both segments have the same resistance. The current flows due to V_{in} and V_o are therefore equal, so they cancel each other. If the potentiometer is set such that $R_{ba} < R_{bb}$, then input current $I1$ is increased due to a decrease of Z1. For frequencies at which Z1 is minimum, the gain is maximum, so the action is to boost those frequencies.

When R_b is rotated towards the output side of the circuit, the condition will be $R_{ba} > R_{bb}$. In this condition, impedance Z2 is reduced. This setting has the effect of reducing the gain at the design frequency. That frequency is the point at which the gain either increases or decreases 3 dB from the 0 dB 'flat zone' reference gain:

$$F_{3\text{ dB}} = \frac{1}{2\pi R1 C1} \tag{6-4}$$

Because the feedback network is symmetrical, boost and cut frequencies are the same. The maximum boost or cut will be:

$$B = \frac{R1 || R_b}{R_a || R1} \tag{6-5}$$

or

$$B = \left[\frac{\dfrac{R1 R_b}{R1 + R_b}}{\dfrac{R1 R_a}{R1 + R_a}} \right] \tag{6-6}$$

EXAMPLE 6-2

Calculate the maximum boost or cut in a high frequency shelf equalizer in which $R1 = 100$ kohms, $R_b = 500$ kohms, and $R_a = 10$ kohms.

Solution

$$B = \left[\frac{\dfrac{R1 R_b}{R1 + R_b}}{\dfrac{R1 R_a}{R1 + R_a}} \right]$$

$$B = \frac{\left[\dfrac{(100 \text{ k}\Omega)(500 \text{ k}\Omega)}{(100 \text{ k}\Omega + 500 \text{ k}\Omega)} \right]}{\left[\dfrac{(10 \text{ k}\Omega)(100 \text{ k}\Omega)}{(10 \text{ k}\Omega + 100 \text{ k}\Omega)} \right]}$$

$$B = \frac{\left[\dfrac{50\,000}{600} \right]}{\left[\dfrac{1000}{110} \right]}$$

$$B = \frac{83.3}{9.09} = 9.2$$

A gain of 9.2 represents a gain of a bit less than 20 dB. ∎

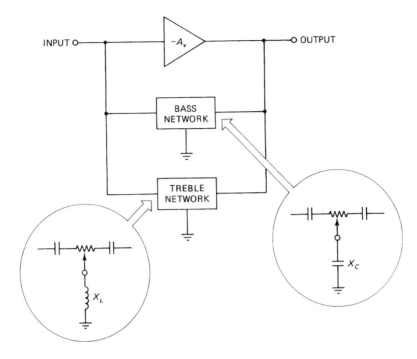

FIGURE 6-18 Reactance tone control circuit.

Other forms of tone control are based on *RL*, *RC* and *RLC* networks in the feedback path. Figure 6-18 shows the block diagram for a type of tone control circuit in which a reactance is used between the wiper of the potentiometer and ground. The bass control uses a capacitive reactance to ground, while the treble control uses an inductive reactance.

The resonant form of tone control is shown in Fig. 6-19. This circuit is similar to the earlier circuit, except that the capacitor between the potentiometer wiper and the op-amp summing junction is a series resonant circuit. A series resonant 'tank' circuit exhibits a very low impedance at the frequency of resonance, while exhibiting a high impedance are frequencies removed from resonance. The resonant frequency of this circuit is found from:

$$F = \frac{1}{2\pi\sqrt{LC}} \qquad (6\text{-}7)$$

The *Baxandall tone control* circuit is shown in Fig. 6-20. The operational amplifier is connected in the inverting follower configuration, but with a different type of feedback network than was used earlier. This type of circuit uses a single bass and treble control, and was once the basis for most earlier high fidelity equipments.

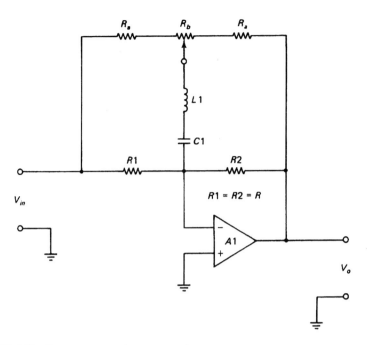

FIGURE 6-19 Resonant network tone control.

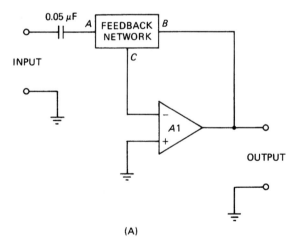

(A)

FIGURE 6-20 Baxandall tone control circuit: (A) overall circuit; (B) feedback network.

198 AUDIO APPLICATIONS OF LINEAR IC DEVICES

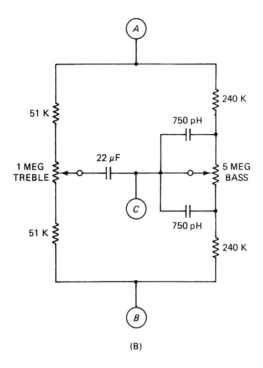

FIGURE 6-20 (continued)

6-7 POWER AMPLIFIERS

The power amplifier in an audio system is designed to boost the signal from a small voltage level up to a power level sufficient to drive a load such as a loudspeaker. While most preamplifiers using small-signal transistors are operated class A, power amplifiers are operated in class B, or class AB. A class A amplifier is one in which the output current flows over all 360° of the input signal cycle. The result of 360° conduction is that the signal is reproduced in its entirety in the output circuit. Class A amplifiers also produce a large amount of heat at any given power level because they are only about 25% efficient. Thus, for a 1 watt output power there will be 3 watts wasted as heat. For this reason, class A amplifiers are only rarely used as audio power amplifiers (some older car radios used class A power amplifiers).

A class B amplifier offers output current conduction only over 180° of the input signal cycle. Thus, the output will be only halfwave. In Fig. 6-21A a class B amplifier with a gain +A is used. The input signal is a sinewave, while the output signal is only half a sinewave. This type of amplifier, if used alone, would result in unacceptable distortion of the input signal. A correction for this problem is the *push-pull power amplifier* shown in Fig. 6-21B. In this

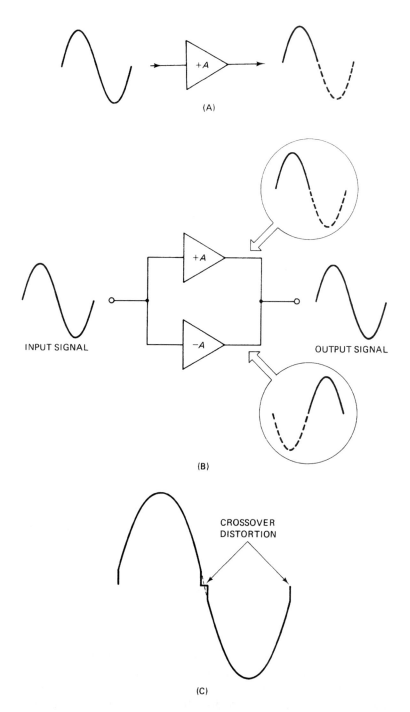

FIGURE 6-21 (A) single-ended amplifier; (B) push-pull amplifier; (C) crossover distortion.

200 AUDIO APPLICATIONS OF LINEAR IC DEVICES

type of audio power amplifier a pair of complementary amplifiers are used, each of which produce the same gain but opposite phase relationship. The $+A$ amplifier is noninverting, and the $-A$ amplifier is an inverting amplifier. If the outputs are combined, then the full sinewave is reproduced.

There is a problem with improperly designed solid-state push-pull amplifiers, however. Figure 6-21C shows *crossover distortion*. This distortion is caused by the 0.7 Vdc junction potential of the transistors being nonlinear. It was crossover distortion that was at the root of the so-called 'transistor sound' imputed to early solid-state high fidelity equipment. Bias arrangements are used to overcome crossover distortion. These amplifiers are called class AB amplifiers because the current flows somewhat more than 180°, but considerably less than 360°.

Linear IC devices tend to be class AB amplifiers, even preamplifiers. The operational amplifier (which forms the basis of most preamplifier circuits) uses one of several class AB circuits in the output stage, even though only producing a few milliwatts of power.

Figure 6-22 shows several implementations of the class AB power amplifier as used in many integrated circuits. The version shown in Fig. 6-22A is called the *complementary symmetry push-pull power amplifier*. It uses a pair of bipolar transistors that are electrically identical except for polarity: one is an NPN and the other is PNP. An NPN transistor turns on harder when base-emitter (b–e) signal goes more positive, and turns off when the b–e signal goes less positive. An NPN transistor, on the other hand, works exactly the opposite: it will turn on harder when the b–e signal goes less positive, and turns off when the b–e signal goes more positive. Thus, NPN transistor $Q1$ will turn on when the input signal is positive, while PNP transistor $Q2$ turns on when the input signal is negative. The transistors are operated as emitter followers, and the output signal is taken at the common junction of the two emitters. The load resistor (e.g. a loudspeaker) forms the emitter resistor for $Q1$ and $Q2$. The voltage across the load tends to be low compared with certain other amplifier circuits.

It is somewhat difficult to build identically matched NPN and PNP power transistors, although somewhat less difficult at smaller signal levels. Because it is easier to build identical transistors of the same polarity on the same silicon substrate many manufacturers use the *totem-pole push-pull amplifier* of Fig. 6-22B. In this circuit a pair of identical NPN transistors are connected in series across the DC power supply.

Because the two transistors are identical, no inherent phase shift exists to accommodate the two halves of the signal. Because of this limitation a phase inverter circuit is used between the input signal and the base terminals of the two power transistors. The phase inverter has two outputs that are 180° out of phase with each other.

If only a signal polarity DC supply is used, then the voltage at the junction of the two transistors will be $(V+)/2$. Because this voltage will interfere with external loads, a large value capacitor (*C*1) is used to block DC from the load. This capacitor typically has a value of at least 1000 µF, and may have a value as high as 10 000 µF. These capacitors are electrolytic types, and are therefore quite large. They also introduce undesirable phase shift to the output signal, especially at lower frequencies. The phase shift is created by

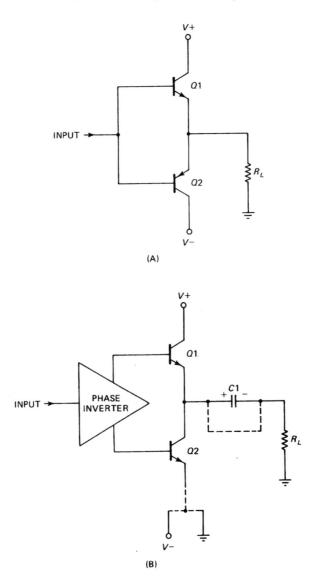

FIGURE 6-22 (A) Push-pull amplifier based on complementary pair power transistors; (B) 'totem-pole' power amplifier circuit gets phase inversion from preamplifier or driver circuit; (C) typical power amplifier circuit.

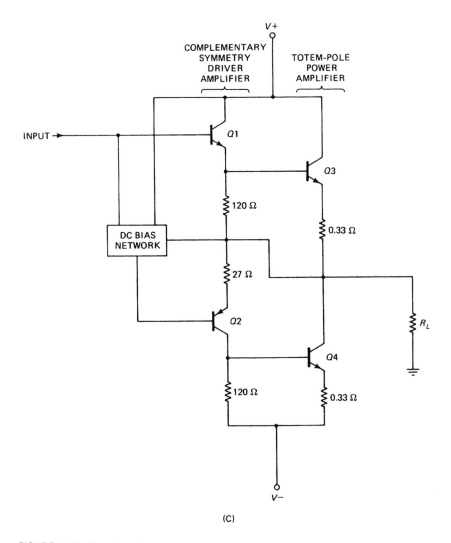

(C)

FIGURE 6-22 (continued)

the capacitive reactance of $C1$. Because of this problem, some totem-pole power amplifiers are designed to use bipolar DC power supplies. Such an arrangement is possible, as in the complementary case, because the positive and negative contributions to the DC level at the output cancel each other.

The circuit that is used in most IC devices, whether power amplifiers or preamplifiers, is the *quasi-complementary power amplifier* shown in Fig. 6-22C. The output section ($Q3/Q4$) is a totem-pole amplifier using a pair of identical NPN transistors. The driver amplifiers are a medium power level transistors connected in a complementary symmetry amplifier circuit. The phase inversion function is handled in this stage.

6-7.1 IC Power amplifiers

The main limitation to producing power amplifiers in IC form was the dissipation of internal heat built up in the process. Class AB amplifiers can be more than 75% efficient (if carefully designed and properly built), but that still leaves considerable heat energy to be dissipated into the environment. But methods were worked out, and IC device power levels started upwards in the mid-1970s. Today, IC and hybrid audio power amplifier devices are available at power levels from 250 mW up to 500 watts. There too many devices on the market to adequately cover them all, so only a pair of representative devices will be examined.

The LM-383 device (Fig. 6-23A) is a 7 watt audio power amplifier packaged in a 5-pin version of the TO-220 plastic power transistor case (see Fig. 6-23B). It will operate at monopolar DC power supply potentials of +5 Vdc to +20 Vdc, and offers a peak current capability of 3.5 amperes. Because this amplifier operates from a monopolar supply, however, there will always be a DC output level of $(V+)/2$ at the output terminals (Fig. 6-24). Similarly, the AC feedback path also requires a DC blocking capacitor.

The LM-383 offers a gain-bandwidth product of 30 kHz at a gain of 40 dB, and a total harmonic distortion (THD) level of 0.2% or less. The series RC network $(C4/R1)$ between the output terminal and ground is a lag compensation circuit designed to prevent oscillation that could occur when high gains are selected.

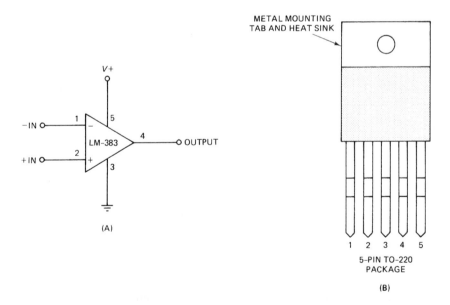

FIGURE 6-23 (A) LM-383 IC power amplifier; (B) package and pinouts.

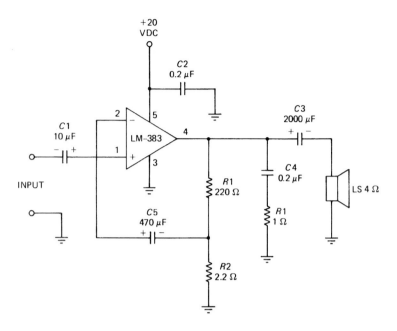

FIGURE 6-24   LM-383 power amplifier circuit.

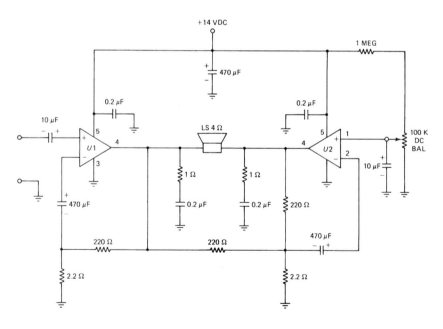

FIGURE 6-25   Bridge amplifier circuit.

AUDIO APPLICATIONS OF LINEAR IC DEVICES   **205**

The circuit of Fig. 6-24 is a single power amplifier configuration for the LM-383 device. It is the circuit that would be used whenever the required power level is within the range of the device. However, there is another way to use these amplifiers is the *bridge amplifier* circuit of Fig. 6-25. This circuit use a pair of LM-383 devices ($U1$ and $U2$) connected such that the load, a loudspeaker, is connected between the two output terminals. One amplifier is connected as a noninverting amplifier ($U1$) while the other is connected as an inverting amplifier ($U2$). The inverting amplifier derives its input signal from the output of $U1$.

No DC blocking capacitors are needed in the output circuits of $U1$ and $U2$ because the DC levels are equal. For example, in a 14.4 Vdc auto radio power amplifier each IC will exhibit +7.2 volts at the output terminal, so the voltage across the load will be $[(+7.2 \text{ Vdc}) - (+7.5 \text{ Vdc})] = 0$. A DC BALANCE control is used at the noninverting input of $U2$ to maintain the zero difference potential over the range of tolerance of the LM-383 devices.

Another example of a power IC device is the Burr-Brown OPA-501 shown in Fig. 6-26. The OPA-501 is not a completely monolithic IC, but rather an

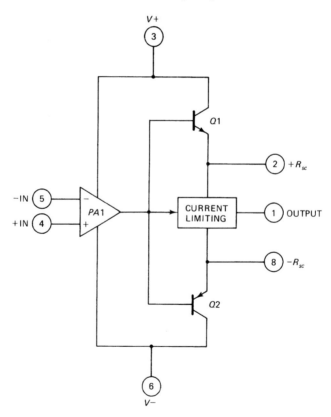

FIGURE 6-26 Internal circuit of the OPA-501.

example of a hybrid. Similar devices in the same family are the OPA-511 and OPA-512 devices. These devices can be used as audio amplifiers, servo amplifiers, motor drivers, synchro drivers and DC power supply regulators. It will operate over a wide range of bipolar DC power supplies: ±10 Vdc to ±30 Vdc.

The OPA-501 uses an operational amplifier for the preamplifier stages. A class AB output stage (Q1/Q2) that will deliver up to 260 watts of peak power and a current of ±10 amperes peak. The OPA-501 is packaged in a 'small size' eight-pin version of the TO–3 power transistor case.

The output current is limited by external circuitry to a level determined by external set-current resistors (see $+R_{sc}$ and $-R_{sc}$ in Fig. 6-27). The value of these two resistors is:

$$R_{sc} = \left[\frac{-0.65}{I_{\text{limit}}} - 0.0437 \right] \text{ ohms} \qquad (6\text{-}8)$$

Current I_{limit} is the maximum output current that the designer wishes to allow. The power dissipation of these resistors is found from:

$$P_d = R_{sc} I_{\text{limit}}^2 \qquad (6\text{-}9)$$

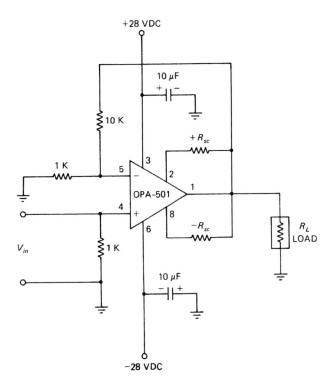

FIGURE 6-27 OPA-501 circuit.

The resistors selected for the limiting function should have a power rating higher than given by Eq. (6-9) by a margin sufficient for reliable and safe operation.

6-8 SUMMARY

1. Audio circuits operate in the range 20 Hz to 20 kHz, although some amplifiers characterized as audio operate over a slightly wider range.
2. Three different forms of audio amplifiers are typically used: *preamplifiers* are (generally) high gain voltage amplifiers, and are used to boost low-level transducer or other input signals to a higher level (usually 100 mV to 500 mV); *control amplifiers* are usually low gain ($\times 1$ to $\times 10$), but perform some signals processing function such as tone control; *power amplifiers* create the audio power level needed to drive the external load.
3. Commercial and broadcast audio amplifiers tend to use 600-ohm input and output impedances. The 600-ohm line is a balanced line that uses to two out of phase signal lines and one ground or common line.
4. Audio mixers linearly combine the signals from two or more sources into one channel. For example, a broadcast studio might mix together signals from two microphones, a tape player and a record turntable.
5. Compression amplifiers offer less gain on signal peaks than on valleys, resulting in an output signal with a higher average power. These devices are used in electronic music applications and in broadcasting.
6. Tone control amplifiers are used to boost or cut specific frequency ranges in order to custom balance the frequency response of the amplifier.
7. Most IC and hybrid power amplifiers use class AB output stages in order to reduce crossover distortion and provide better efficiency (hence less heat dissipation).

6-9 RECAPITULATION

Now return to the objectives and Pre-quiz questions at the beginning of the chapter and see how well you can answer them. If you cannot answer certain questions, place a check mark to each and review the appropriate parts of the text. Next, try to answer the questions and work the problems below, using the same procedure.

6-10 STUDENT EXERCISES

1. Design, build and test a noninverting audio amplifier with a frequency response of 300 Hz to 3000 Hz, using an operational amplifier.
2. Design, build and test a shelf equalizer (Fig. 6-17A) with a center frequency of 1500 Hz.
3. Design, build and test an audio amplifier based on an LM-381.

6-11 QUESTIONS AND PROBLEMS

1. What is the nominal 'audio' frequency range?
2. List four applications for audio amplifiers.
3. Draw the circuit for a bridge audio power amplifier based on the LM-383 integrated circuit power amplifier. Is it necessary to use output DC blocking capacitors?
4. An audio preamplifier for an AM broadcast radio transmitter must have a gain of 100, and a frequency response of 30 Hz to 5000 Hz between −3 dB points. Calculate the values for the input and feedback resistors, and the value of the capacitor shunting the feedback resistor. Assume a noninverting follower.
5. In the circuit above, the input circuit contains a transformer with a 1:2 turns ratio. What is the total voltage gain of this circuit?
6. Draw the circuit for a five-channel op-amp audio mixer with an inverting characteristic.
7. A circuit is used to boost the 5 mV output of a dynamic microphone to 0.100 volt needed to drive the input of a radio transmitter. How much gain is required? Draw a noninverting follower that has this amount of voltage gain.
8. An inverting audio amplifier is built from an operational amplifier with a 150 kohm resistor in the feedback network. Calculate the value of capacitance that must be shunted across the feedback resistor in order to limit the upper −3 dB point in the frequency response to 4 kHz.
9. In an inverting amplifier the feedback resistor is 220 kohms, and it is shunted with a 100 pF capacitor. Calculate the frequency at which the gain drops −3 dB.
10. A four-input mixer circuit must be designed and built. Assume a feedback resistance of 150 kohms and that each input has a 100 kohm resistance. Calculate the total output voltage if the following DC voltages are applied, $V1 = 0.1$ Vdc, $V2 = 0.2$ Vdc, $V3 = 0.1$ Vdc, and $V4 = 0.4$ Vdc. Calculate the peak output voltage if the DC sources are replaced with the following rms signal voltages, $V1 = 0.05$ V_{rms}, $V2 = -0.1$ V_{rms}, $V3 = -0.43$ V_{rms} and $V4 = 0.23$ V_{rms}.
11. The simplest form of tone control is the _____ _____ .
12. Draw the circuit diagram for a shelf equalizer stage (Fig. 6-17A) in which $R1 = R2 = 100$ kohms, $R \approx 10$ kohms, and $R'' = 500$ kohms. Calculate the capacitance needed for $C1$ for a 1500 Hz 3 dB point. Calculate the maximum boost or cut for this circuit (in dB).
13. A resonant equalizer tone control (Fig. 6-19) requires a resonant frequency of 2000 Hz. Calculate the capacitance required if the inductor is a 900 mH unit.
14. What is the purpose of $C6$ in Fig. 6-2?
15. What is the input impedance and output impedance of the circuit in Fig. 6-2?
16. Draw the frequency response curve of Fig. 6-2.
17. Sketch the frequency response curves of Fig. 6-4 when the feedback network is (a) Fig. 6-4C, and (b) Fig. 6-4D.

18. In Fig. 6-11A the transformer has a turns ratio of 1:3, and both resistors in the feedback network are equal to each other. What is the peak-to-peak output voltage when a 100 mV$_{rms}$, 1000 Hz sinewave signal is applied to the input?
19. Select a value for $R3$ and $R4$ in Fig. 6-11B when the overall differential gain must be 100.
20. Sketch the circuit diagram for a transformerless 600-ohm line driver amplifier based on op-amps.
21. Sketch the circuit for a Baxandall tone control circuit.
22. A amplifier reduces its gain on input signal peaks and increases the gain on signal valleys.
23. An audio power amplifier drives an 8 ohm resistance load. A potential of 22.4 volts peak-to-peak is measured across the load when a 1000 Hz sinewave is applied to the input of the amplifier. (a) Calculate the output power; (b) calculate the amplifier efficiency if the DC power consumption of the stage is 34 watts.
24. An audio power amplifier produces a maximum output power of 100 watts into a 4 ohm resistive load. What is the DC power drawn by this amplifier if it is 33% efficient?
25. What is the cause of crossover distortion? What is the usual circuit fix?

CHAPTER 7

Communications applications of linear IC devices

OBJECTIVES

1. Learn the types of applications for linear ICs in communications circuits.
2. Understand how to use phase locked loop IC devices.
3. Understand how to use the analog multiplier/divider.
4. Learn how to use current loop systems.

7-1 PRE-QUIZ

These questions test your prior knowledge of the material in this chapter. Try answering them before you read the chapter. Look for the answers (especially those you answered incorrectly) as you read the text. After you have finished studying the chapter try answering these questions again, and those at the end of the chapter (see Section 7-10).

1. List two applications for current loop communications.
2. List the voltage levels that represent logical 1 and logical 0 in the RS-232C standard.
3. Amplitude modulation is a _____ process.
4. The _____ signal is a form of amplitude modulation in which the carrier is suppressed, and one sideband is filtered out.

7-2 INTRODUCTION

The broad area of 'communications' takes in a wide range of technology, only a small portion of which can be addressed in a single chapter. Those areas that we can touch on, however, are nonetheless varied and broad because they relate to the applications of linear IC devices in communications. In this chapter a number of different areas are considered as examples: *current loop communications, serial voltage level communications, modulation and demodulation, and phase locked loops*.

7-3 CURRENT LOOP SERIAL COMMUNICATIONS

Current loop communications is one of the oldest form of machine communication known. Applications of the current loop include teletypewriters and process control instrumentation (a varied field in itself). The current loop offers some advantages even today, despite the fact that it is seemingly old-fashioned (and is largely replaced by other methods). First, the current loop is somewhat less sensitive to noise problems brought on by voltage droops or spikes on the lines. Second, it is relatively easy to 'daisy chain' several instruments into one system. Second, the current loop can be implemented using a simple 'twisted pair' of wires rather than a multi-wire cable or a coaxial transmission line as are required in other (faster) forms of data communications.

A disadvantage is that a single point failure can take the entire system down. If a single load on the series current loop opens up, then the entire chain is inoperative (although not all failures in series loop machines will cause this problem). In this text two forms of current loop will be examined: *teletypewriters* (Section 7-3.1) and *process instrumentation* (Section 7-3.2).

7-3.1 Teletypewriter current loops

One of the earliest forms of alphanumeric data communications was the teletypewriter machine. These were typewriter-like (or printer-like) machines that used a mechanism of electrical solenoids to pull in the type bars, or, to position the type printing cylinder (which struck an inked ribbon). The original devices used the now obsolete 5-bit BAUDOT code, and a 60 mA current loop. Later versions of the teletypewriter machine used the modern 7-bit ASCII code (still in common use) and a 20 mA current loop. These machines used the same basic technique as earlier machines, but were generally more sophisticated than previous designs. A few modern teletypewriters use dot matrix printing and contain a floppy disk to store a magnetic copy of the data transmitted and received, yet nonetheless operate from the 20 mA current loop.

Figure 7-1 shows the basic elements of a teletypewriter (or other printer) based on the 20 milliampere current loop. The keyboard and printer are actually separate units, and they usually have to be wired together if a local loop is desired (i.e. where the keystroke on the keyboard produces a printed character on the same machine). This circuit is simplified. In a real teletypewriter there will be an encoder wheel or circuit that produces the binary coded output. The keyboard consists of a series of switches (that actuate the encoder). Since these switches and their associated encoder are in series with the line, a 'LOCAL' switch must be provided to bypass the transmitter section on receive.

The receiver consists of a decoder and the receive solenoids, which operate the typebar mechanism. Note in Fig. 7-1 that a 1N4007 diode is shunted across the receive solenoid. This diode is used to suppress the high voltage inductive spike that is generated when the reactive solenoids are de-energized. The diode is placed in the circuit such that it will be reverse biased under normal operation. But the transient counter electromotive force (CEMF) produced as a result of 'inductive kick' briefly forward biases the diode. The diode therefore clamps the high voltage spike to a harmless level. In some older machines the inductive spike was safely ignored because the mass of

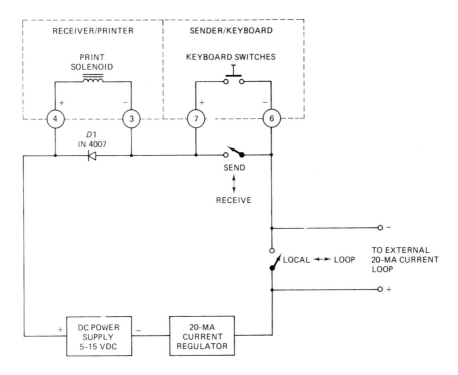

FIGURE 7-1 Current loop teletypewriter control.

COMMUNICATIONS APPLICATIONS OF LINEAR IC DEVICES   **213**

the mechanism effectively integrated the spike to nothingness. But modern solid-state equipment does not move the mechanism directly with the 20 mA loop, but rather through electronic drivers. The solid-state components can be damaged by the high voltage spike, so it is recommended that a 1N4007 be used even if the original design ignored it.

When the loop is closed, the circuit of Fig. 7-1 will produce a readable signal. Another similarly-designed teletypewriter will be able to read the current variations produced by the machine. The binary '1' (HIGH) condition, also called a 'MARK' in teletypewriter terminology, is defined as a current of 16 mA to 20 mA. The binary '0' (LOW), also called a 'SPACE', is defined as a current between 0 mA and 2 mA.

Figures 7-2 and 7-3 shows how to interface 20 mA current loop equipment to TTL-compatible serial outputs on digital computers. The circuit in Figure 7-2 shows the transmitter arrangement. The assumption is that there is a single TTL-compatible bit from either a serialized-parallel output, or, a UART IC. The TTL level is applied to an open collector TTL inverter, which has as its collector load an LED inside of an IC optoisolator. When the LED is turned on, the phototransistor is turned on hard. This transistor operates as an electronic switch in series with the 20 mA current loop. Thus,

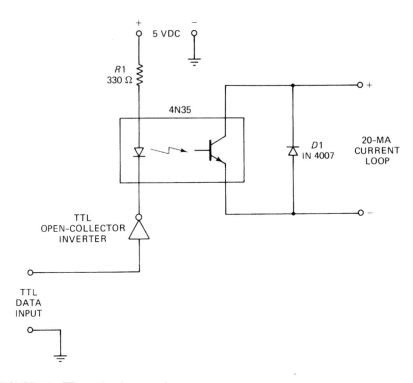

FIGURE 7-2 TTL-to-20 mA current loop converter.

214 COMMUNICATIONS APPLICATIONS OF LINEAR IC DEVICES

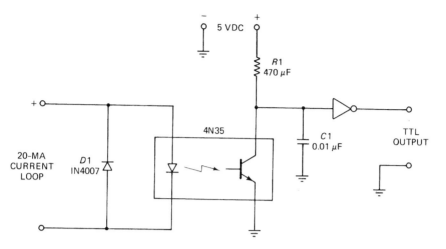

FIGURE 7-3   20 mA current loop to TTL converter.

when the TTL bit is HIGH, the LED is on and the transistor is saturated. In that condition, the current loop transmited a 'MARK' sign (equivalent to a logical 1 in binary).

The receive end of the current loop-to-TTL interface is shown in Fig. 7-3. In this case, the optoisolator is still used, but in reverse. Here the LED is connected in series with the current loop. Thus, when a MARK is transmitted, the LED will be turned on; when a SPACE is transmitted the LED is turned off. During the MARK periods, the optoisolator phototransistor is saturated, and the input to the TTL inverter is LOW. This condition results in a HIGH on the output to the computer. Again, a MARK is a logical 1 (HIGH) and a SPACE is a logical 0 (LOW). The 0.01 µF capacitor is used for noise suppression.

7-3.2   4-to-20 Milliampere current loops

Industrial process control technology uses a current loop system in some instrumentation communications applications. Figure 7-4A shows a typical system in which three different devices are served by the same current loop. The current has a range of permissible values of 4 mA to 20 mA (see Fig. 7-4B). The purpose of the current loop is to transmit a range of values that represent parameters being measured. The overall dynamic range of the current is 16 mA, but the 4 mA offset raises the maximum current to 20 mA. The reason for the offset, and one of the advantages of the 4-to-20 mA instrumentation loop over the teletypewriter loop, is that the zero input condition is represented by a known current (4 mA), not 0 mA. Thus, the 4-to-20 mA loop can distinguish between a zero parameter value (represented by 4 mA) and a zero current condition brought on by a failure in the

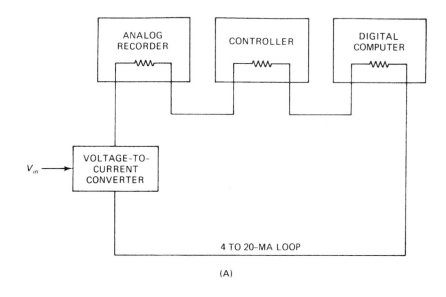

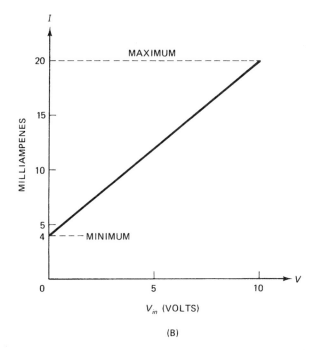

FIGURE 7-4 (A) daisy chaining 4-to-20 mA circuit; (B) transfer function.

circuitry or an open communications line, or the fact that the equipment is turned off.

There is sometimes some unnecessary mystification of the 4-to-20 mA current loop. Some students ask 'What does the range represent?' It doesn't represent anything in general, it's just an available dynamic range that can be appropriated to represent anything desired. For example, in Fig. 7-4B the 4-to-20 mA current range is used to represent a 0 to 10 volt signal range. Thus, when the signal is 0 volts a 4 mA current is transmitted, while a +10 volt signal produces a 20 mA current. The designer can redefine the meanings of the 4 mA minima current and the 20 mA maxima current according to the needs of the system.

A voltage-to-current converter is needed to make the current loop work with ordinary input devices. Unlike the teletypewriter case, where the current was either on or off depending upon whether a '1' or '0' was being transmitted, the analog 4-to-20 mA current loop requires a circuit that will produce an output load current that is proportional to the input voltage. Figure 7-5 shows one such circuit. The current loop, here represented by load resistor R_L, is the feedback current of a noninverting follower operational amplifier circuit. The current is proportional to the input voltage according to the following rule:

$$I_L = \frac{V_o - V_{in}}{R_L} \qquad (7\text{-}1)$$

Output voltage V_o is found from:

$$V_o = \frac{V_{in} R_L}{R_{in}} + V_{in} \qquad (7\text{-}2)$$

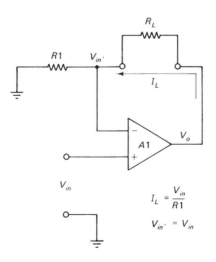

FIGURE 7-5 Voltage-to-current converter.

By algebra one can demonstrate that substituting Eq. (7-2) into Eq. (7-1) produces:

$$I_L = \frac{V_{in}}{R1} \tag{7-3}$$

The circuit of Fig. 7-5 operates over a wide range of load resistance values, but suffers from the fact that the load must be floating with respect to ground. A grounded load circuit is shown in Fig. 7-6. This circuit is called the *Howland current pump* after B. Howland of the MIT Lincoln Laboratories in Massachusetts, USA. The uniqueness of the Howland current pump is that the load current is independent of load resistance. The current is proportional to the difference between voltages $V1$ and $V2$. In the generalized case of the Howland current pump, $V2$ is not zero, although in many practical applications $V2$ is zero and the associated input resistor is grounded. Figure 7-7 is a graphical solution of the circuit evaluation for the case $V2 = 0$.

Another distinction of the Howland current pump is that it can either sink or source current depending upon the relationship of $V1$ and $V2$. If $I_L > 0$, then the current flows out of the circuit (i.e. 'down' in Fig. 7-6), but if $I_L < 0$

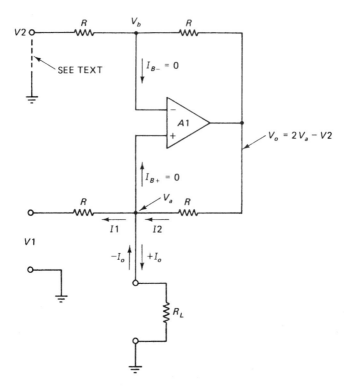

FIGURE 7-6 Howland current pump circuit.

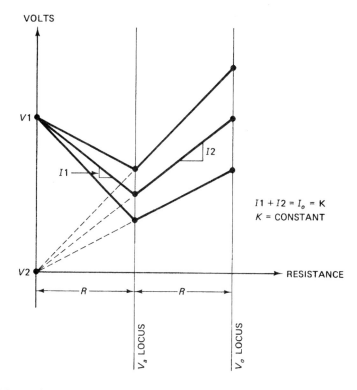

FIGURE 7-7 Graphical solution of current converter.

the current flows into the circuit (i.e. 'up' in Fig. 7-6). For sake of simplicity the case where $V2 = 0$ is evaluated.

There are two 'output' voltages in this circuit. Voltage V_a is the load voltage and appears across the load resistance:

$$V_a = I_L R_L \tag{7-4}$$

According to Kirchhoff's current law (KCL):

$$I_L = I1 + I2 \tag{7-5}$$

By Ohm's law:

$$I1 = \frac{V1 - V_a}{R} \tag{7-6}$$

and

$$I2 = \frac{V_o - V_a}{R} \tag{7-7}$$

COMMUNICATIONS APPLICATIONS OF LINEAR IC DEVICES **219**

Substituting Eqs (7-6) and (7-7) into Eq. (7-5):

$$I_L = \frac{V1 - V_a}{R} + \frac{V_o - V_a}{R} \tag{7-8}$$

which reduces to:

$$I_L R = V1 + V_o - 2V_a \tag{7-9}$$

Therefore, we may write:

$$V_a = \frac{V1 + V_o - I_L R}{2} \tag{7-10}$$

For voltage V_a the operational amplifier is connected in the noninverting follower configuration. Because the two feedback network resistors are equal, the gain of the circuit is two. Therefore:

$$V_o = 2V_a \tag{7-11}$$

or

$$V_o = V1 + V_a - I_L R \tag{7-12}$$

Combining terms:

$$0 = V1 - I_L R \tag{7-13}$$

$$I_L = \frac{V1}{R} \tag{7-14}$$

Equation (7-14) informs us that output load current I_L is proportional to input voltage V_{in} and inversely proportional to the feedback resistance (assuming all four resistors are equal).

In the more general case where $V2$ is non-zero, Eq. (7-14) becomes:

$$I_L = \frac{V1 - V2}{R} \tag{7-15}$$

To make the Howland current pump produce the 4 mA current that represents the 0 volts signal level requires that $V2$ be set to a value that will force Eq. (7-15) to evaluate to 4 mA when $V1 = 0$.

EXAMPLE 7-1

A Howland current pump is to be used to make a 4-to-20 mA current loop transmitter. What value of bias voltage $V2$ will force I_L to be 4 mA when $V1 = 0$; assume $R = 1$ kohm.

Solution

$$I_L = \frac{V1 - V2}{R}$$

$$4 \text{ mA} = \frac{0 \text{ volts} - V2}{1 \text{ k}\Omega}$$

$$4 \text{ volts} = 0 - V2$$

$$V2 = -4 \text{ volts} \qquad \blacksquare$$

Now that the current loop has been examined, let's turn our attention to other serial communications methods in which the signal is a voltage level.

7-4 RS-232 SERIAL INTERFACING

Serial data communications requires only one channel (i.e. either a pair of wires, or a single radio or telephone channel), so is ultimately less costly than parallel data transmission. The benefit is especially noticeable on long line systems where the extra cost of wire and/or telecommunications channels becomes most apparent. The current loops discussed in the previous section were the earliest form of interface to peripherals, and are still in use (albeit declining in popularity). In this chapter, we will discuss what is probably the most common form of voltage operated serial communications channel, the RS-232C port.

The RS-232C standard concerns itself with serial data transmission using voltage levels to represent '1' and '0', rather than current levels. The RS-232C

TABLE 7-1 RS-232 Pin assignments.

Pin No.	RS-232 Name	Function
1	AA	Chassis ground/common
2	BA	Data from terminal
3	BB	Data received from MODEM
4	CA	Request to send
5	CB	Clear to send
6	CC	Data set ready
7	AB	Signal ground
8	CF	Carrier detection
9	(x)	
10	(x)	
11	(x)	
12	(x)	
13	(x)	
14	(x)	
15	DB	Transmitted bit clock (internal)
16	(x)	
17	DD	Received clock bit
18	(x)	
19	(x)	
20	CD	Data terminal ready
21	(x)	
22	CE	Ring indicator
23	(x)	
24	DA	Transmitted bit clock, external
25	(x)	

(x) = unassigned

standard calls for the use of a 25-pin D-shell standard connector (the type DB-25) that is always wired in exactly the same manner (see Table 7-1) and using the same voltage levels. If all signals are defined according to the standard, then it is possible to interface any two RS-232C devices without any problems.

A large collection of peripherals (modems, printers, video terminals) are fitted with DB-25 RS-232C connectors. Unfortunately, there are some manufactures who also use the DB-25 series of connectors in exactly the same gender as RS-232C, but not in the RS-232C manner. Thus, the mere existence of an RS-232C connector is not adequate proof of RS-232C in use.

In recent times, the personal computer industry has created a nine-pin subset of the RS-232C standard using a DB-9 series connector that is similar to the DB-25 except for size and the number of pins.

The RS-232C standard is older than most of our present-day digital devices, so uses an obsolete set of voltages for the levels. In RS-232C format, the logical 1 (HIGH) is represented by a potential between -5 and -15 volts, while a logical 0 (LOW) is represented by a potential between $+5$ and $+15$ volts. Since most digital equipments today are based on TTL-compatible formats ($0 = 0$ volts, $1 \geq +2.4$ volts), some level translation is needed. Perhaps the most common conversion method is to use the 1488 line driver and 1489 line receiver chips.

The circuit of Fig. 7-8 is used to convert a 20 mA current loop signal, as might be found with older design teletypewriters or printers, into an RS-232C signal. The circuit is designed around an optoisolator ($U1$). The optoisolator is an IC device in which light from an LED shines on the base region of a phototransistor. This feature is especially desirable if the current loop experiences high voltage transient spikes from the 'inductive kick' of print solenoids. The output of the optoisolator is connected to the noninverting input of an operational amplifier ($A1$); the inverting input is biased to $+5$ Vdc. When the LED is turned off (e.g. during the '0' condition on the loop), the transistor is also turned off so the voltage at point A is high (essentially $V+$). This voltage applied to the noninverting input means that the differential input voltage is effectively negative, which forces the output of $A1$ to saturate at $+V_o$ (about $+12$ Vdc).

The LED in the optoisolator is turned on when the 20 mA current flows. Resistor $R1$ limits the current in the LED to a safe value within overburdening the loop with an excessive voltage drop. The 20 mA current turns on the LED, which in turn illuminate the phototransistor ($Q1$) and drives it into saturation. When this occurs the voltage on the collector (point A) drops to near zero. Under this condition the differential input voltage of the op-amp is positive, so the output saturates to $-V_o$, or about -12 Vdc.

The circuit shown in Fig. 7-9 is used to convert TTL levels from a computer to RS-232C levels for transmission. This circuit is also based on the operational amplifier ($A1$). The -15 Vdc and $+15$ Vdc power supplies

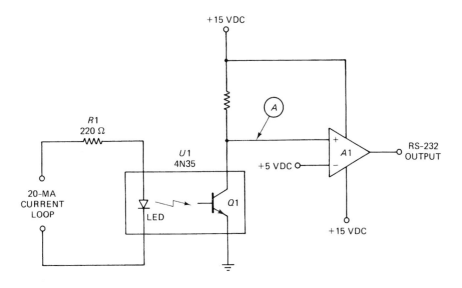

FIGURE 7-8 20 mA to RS-232 converter.

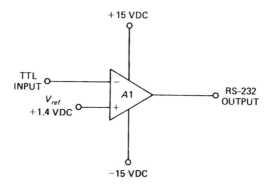

FIGURE 7-9 TTL-to-RS-232 converter.

used for the operational amplifier are the source of the RS-232C levels. The TTL input is biased for noise band immunity by a reference voltage (V_{ref}) of 1.4 Vdc. The TTL levels are: HIGH ('1') = +2.4 to +5 volts, while LOW ('0') = 0 to 0.8 volts. If a '1' is applied to the input of Fig. 7-9, then the differential input voltage (+2.4 − 1.4, or +1 volt) is positive, so the output

COMMUNICATIONS APPLICATIONS OF LINEAR IC DEVICES   **223**

of A1 is forced to $-V_o$. Similarly, when the input is a '0', then the differential input voltage of A1 is $0 - 1.4$, or -1.4 volts, so the output is forced to $+V_o$.

7-5 MODULATION AND DEMODULATION

Modulation is the process of using one signal to alter a characteristic of another signal in order to convey information. In the context of communications, the altered signal is called the carrier signal, while the other signal is called the modulating signal. The carrier signal usually has a much higher frequency than the modulating signal. For example, in an AM band radio signal the carrier will have a frequency between 550 kHz and 1620 kHz, while the audio modulating signal will have a frequency less than 5 kHz. Alternatively, in a wire carrier telephony system, the carrier might have any frequency between 5 kHz and 200 kHz, while the modulating signal is limited to the range 300 Hz to 3000 Hz.

Any of several carrier parameters can be altered by the modulating signal. In an *amplitude modulation* (AM) system the modulating signal varies the amplitude of the carrier signal. There are also two forms of angular modulation: *phase modulation* (PM) and *frequency modulation*. In these systems the carrier amplitude remains constant, but either its frequency or phase is varied. The two forms of angular modulation are similar enough to each other that the terms are sometimes used interchangeably. For example, many 'FM' transmitters used in landmobile radios are actually PM devices. Although interchanging 'PM' and 'FM' works in most practical situations, it is technically an error.

AM, PM and FM usually involve a sinusoidal modulating signal altering a sinusoidal carrier. If the carrier is a squarewave or pulse train, however, we can use *pulse amplitude modulation* (PAM), *pulse width modulation* (PWM) and *pulse position modulation* (PPM).

7-5.1 Amplitude modulation (AM)

In AM systems a low frequency modulating frequency (with frequency f_m and amplitude v_m) is used to alter the amplitude of a higher frequency carrier signal (frequency f_c at amplitude v_c). These signals are defined as follows for the sinusoidal case:

$$v_m = V_m \sin(2\pi f_m t) \qquad (7\text{-}16)$$

and

$$v_c = V_c \sin(2\pi f_c t) \qquad (7\text{-}17)$$

where:
    v_m is the instantaneous amplitude of the modulating signal
    V_m is the peak amplitude of the modulating signal
    v_c is the instantaneous amplitude of the carrier signal

V_c is the peak amplitude of the carrier signal
f_c is the carrier frequency in hertz (Hz)
f_m is the modulating frequency in hertz (Hz)
t is time in seconds

Before examining amplitude modulation let's first discuss what AM is not. Simply combining the two signal, v_m and v_c, together in a linear network does not produce AM. Figure 7-10A shows v_m and v_c in separate traces. In Fig. 7-10B the two signals are combined together in a linear resistor network. The result is the *additive* waveform shown in Fig. 7-10B. The instantaneous amplitude of this signal is given by:

$$v = v_m + v_c \tag{7-18}$$

$$v = V_m \sin(2\pi f_m t) + V_c \sin(2\pi f_c t) \tag{7-19}$$

Compare Fig. 7-10B with Fig. 7-10C. The waveform in Fig. 7-10C is a sinewave carrier amplitude modulated with lower frequency sinewave. The AM process is *multiplicative*, so the instantaneous amplitude is:

$$v = (V_c + v_m)\sin(2\pi f_c t) \tag{7-20}$$

The instantaneous envelope is expressed in the form:

$$v_e = V_c + v_m \tag{7-21}$$

or, to rewrite Eq. (7-20) in terms of the envelope expression:

$$v = v_e \sin(2\pi f_c t) \tag{7-22}$$

and, accounting for v_m:

$$v = (V_c + V\sin(2\pi f_m t))(\sin(2\pi f_c t)) \tag{7-23}$$

The ratio of modulating signal to carrier signal is called the *modulation index* (m):

$$m = \frac{V_m}{V_c} \tag{7-24}$$

[NOTE: $0 < m \le 1$]
Equation (7-23) can be expressed in terms of m:

$$v = V_c[1 + m\sin(2\pi f_m t)(\sin(2\pi f_c))] \tag{7-25}$$

or, when normalized:

$$v = [1 + m\sin(2\pi f_m t)(\sin(2\pi f_c t))] \tag{7-26}$$

$$v = \sin(2\pi f_c t) + [(m\sin(2\pi f_m t)(2\pi f_c t)] \tag{7-27}$$

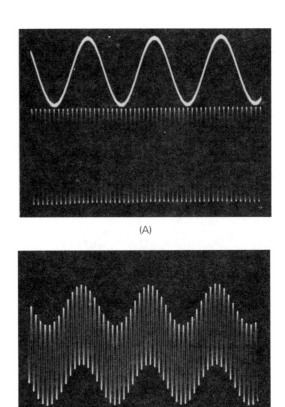

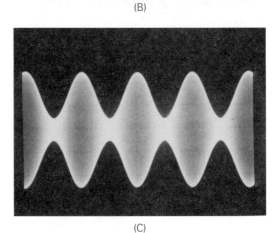

FIGURE 7-10 (A) Low frequency modulating signal and high frequency carrier signal; (B) linear mixing; (C) amplitude modulation.

A well-known trigonometric identity allows us to gain some insight into the nature of amplitude modulation as derived from Eq. (7-27) and the preceding mathematical argument:

$$\cos(A \pm B) = \cos(A)\cos(B) \mp \sin(A)\sin(B) \qquad (7\text{-}28)$$

Applying the identity of Eq. (7-28) to Eq. (7-27) will result in:

$$v = \sin(2\pi f_c t) + \cdots + \left[\frac{m}{2}\right][\cos(2\pi f_c - 2\pi f_m)t - \cos(2\pi f_c + 2\pi f_m)t] \qquad (7\text{-}29)$$

Notice the terms $(2\pi f_c - 2\pi f_m)$ and $(2\pi f_c + 2\pi f_m)$ in Eq. (7-26). These terms tell us that sum and difference frequencies are generated. The sum frequency is called the *upper sideband* (USB) and the difference frequency is called the *lower sideband* (LSB). Figure 7-11 shows the result of amplitude

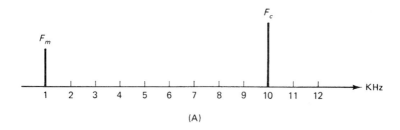

(A)

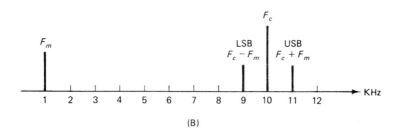

(B)

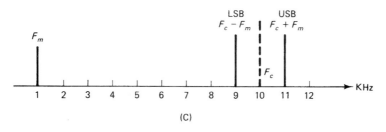

(C)

FIGURE 7-11 (A) Carrier and single-tone modulating frequency; (B) amplitude modulated signal spectrum; (C) suppressed carrier AM.

modulation on the frequency domain as two separate spikes. In Fig. 7-11A the carrier (f_c) and modulating (f_m) signals are shown in frequency domain as two spikes. In Fig. 7-11B the result of amplitude modulation is shown. The carrier frequency is 10 kHz, while the modulating frequency is 1 kHz. The USB is $(10 + 1)$ kHz, or 11 kHz, while the LSB is $(10 - 1)$ kHz, or 9 kHz. In other words, the sidebands vary from the carrier frequency by an amount equal to the modulation frequency. Figure 7-12A shows a simple modulator that is based on a single PN junction diode and a resistor combining network. The small-signal diode provides the nonlinearity over cyclic excursions of the input signal which is required to produce AM. In Fig. 7-12B the result of using a 1 kHz triangle waveform to modulate a 5 kHz squarewave is shown. The modulating signal is superimposed on the carrier in a manner quite different from the linear mixing seen previously in Fig. 7-10B. Another form of simple AM modulator is shown in Fig. 7-13A, with the resultant waveform shown in Fig. 7-13B. Again a triangle wave was used to modulate a higher frequency squarewave signal. When the carrier signal is modulated by an active modulator the result is the bipolar waveform seen in Fig. 7-10C. A point for point explanation of this type of waveform is shown in Fig. 7-13C, but with a triangle modulating signal and a squarewave carrier (for simplicity of illustration).

In Fig. 7-13A the modulation effect is obtained from a CMOS or MOS electronic switch shunted across the signal line. A high-pass filter passes only the carrier and sidebands to the output circuit. The modulating signal is passed along the line, but it is chopped at the carrier frequency because the control terminal on the switch is driven by the carrier signal, v_c. Demodulation of this signal is through a similar process, but with a low-pass filter instead (see Fig. 7-13C). This process is called synchronous demodulation. Another form of demodulation is *asynchronous demodulation* or *envelope detection*. This

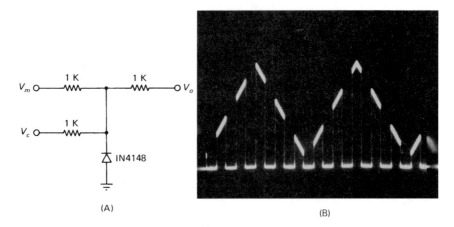

FIGURE 7-12 (A) simple diode amplitude modulator; (B) output waveform.

method uses a rectifier in series with the signal line followed by a low-pass filter circuit.

One final form of amplitude modulator is the analog multiplier shown in Fig. 7-14. Recall from above that the AM signal is the product of the carrier and modulating signals. Thus, it is suitable to use an analog multiplier circuit AM. More will be said about these circuits in Chapter 8. The output of this circuit is:

$$V_o = \frac{V_m V_c}{10} \quad (7\text{-}30)$$

A circuit called a *balanced modulator* is used to produce an output signal that reduces or suppresses the carrier signal, while retaining the sidebands. The spectrum shown in Fig. 7-10C represents this type of signal. It is frequently used in commercial, amateur and military radio communications. The carrier signal (f_c) is suppressed a large amount (ideally to zero, but more likely −40 to −60 dB of suppression), leaving only the upper and lower sidebands. The advantage of this system is redistribution of the available power to only the sidebands. This form of amplitude modulation is usually

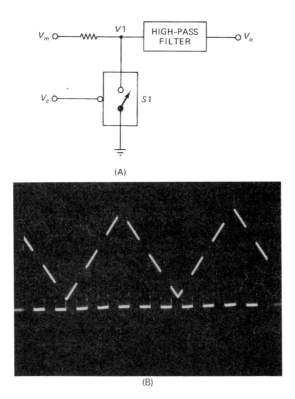

FIGURE 7-13 (A) switch amplitude modulator; (B) output waveform; (C) timing waveforms; (D) detector circuit.

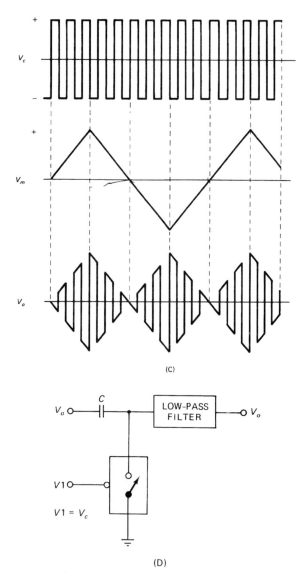

(C)

(D)

FIGURE 7-13 (*continued*)

FIGURE 7-14 Analogue X-Y multiplier.

called *double sideband suppressed carrier* (DSBSC). A related, and more common, form of AM is *single sideband suppressed carrier* (SSBSC, or simply SSB). The suppression of the carrier in DSBSC, and in most forms of SSBSC, is the result of the balanced modulator. Consider the following:

$$v = v_m \times V_c \tag{7-31}$$

$$v = V_m \sin(2\pi f_m t) \times V_c \sin(2\pi f_c t) \tag{7-32}$$

$$v = V_m V_c [\sin(2\pi f_m t)][\sin(2\pi f_c t)] \tag{7-33}$$

Another form of Eq. (7-32) is sometimes seen in the literature because it better reflects the situation in real AM systems. The other form of the AM equation takes advantage of a standard trigonometric identity:

$$\sin(A)\sin(B) = \frac{\cos(A-B) - \cos(A+B)}{2} \tag{7-34}$$

Let:
$$A = V_c = V_{cp} \sin(2\pi f_c t) \tag{7-35}$$

and
$$B = V_m = V_{mp} \sin(2\pi f_c t) \tag{7-36}$$

Substituting Eqs (7-24) and (7-25) into Eq. (7-23):

$$V = \frac{1}{2} V_{mp} V_{cp} \cos[2\pi(f_c - f_m)] - \frac{1}{2} V_{mp} V_{cp} \cos[2\pi(f_c + f_m)] \tag{7-37}$$

Compare Eq. (7-36) to Eq. (7-29). Note that the carrier term $\sin(2\pi f_c t)$ is missing, indicating that the carrier is suppressed. The signal of Eq. (7-36) is a DSBSC signal. If filtering or phasing is used to remove either upper or lower sideband, then the signal becomes a single-sideband (SSB) signal. In communications, the SSB signal is probably the most widely used form of AM except in the VHF aviation band and 27 MHz citizen's band.

The SSB or DSBSC signal cannot be demodulated in the same manner as a standard AM signal. If an envelope detector is used, then the recovered modulating signal will be distorted beyond recognition. The SSB and DSBSC signals are demodulated in a *product detector*. That is, a detector circuit that nonlinearly mixes a signal that represents a reconstructed version of the carrier with the DSB or SSB modulated signal, to recover the modulation. For example, if a 455 kHz carrier is modulated with a 1 kHz sinewave in a balanced modulator, the two signals produced will be the difference (454 kHz) and sum (456 kHz). Only one of these need be transmitted. If the transmitted signal is then mixed with another 455 kHz carrier at the receiver end it will produce the difference frequencies: 456 kHz − 455 kHz = 1 kHz, and 455 kHz − 454 kHz = 1 kHz. The modulation is thus recovered.

Figure 7-15A shows an IC balanced modulator, the LM-1496. The 14-pin DIP package is shown in Fig. 7-15B, while the 10-pin metal can package

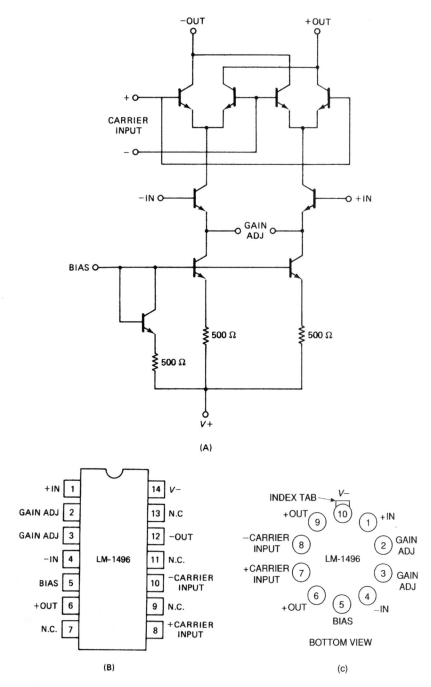

FIGURE 7-15 (A) LM-1496 internal circuit; (B) package and pinouts for DIP case; (C) package and pinouts for metal can.

232 COMMUNICATIONS APPLICATIONS OF LINEAR IC DEVICES

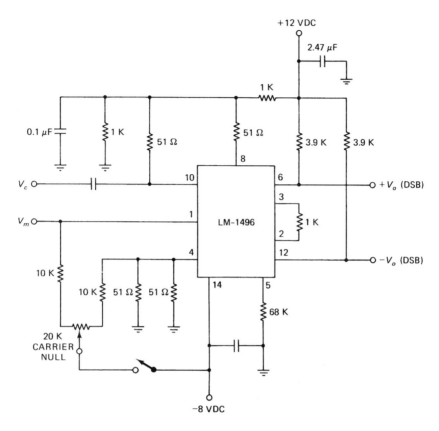

FIGURE 7-16 LM-1496 modulator circuit.

is shown in Fig. 7-15C. These IC balanced modulators will suppress the carrier 65 dB at 500 kHz and 50 dB at 10 MHz. The circuit for a balanced modulator based on the LM-1496 and LM-1596 is shown in Fig. 7-16A. The carrier signal is applied to the +CARRIER input (pin no. 8), while the signal is applied to the +IN input (pin no. 1). The −IN and −CARRIER inputs are terminated. The optimum signal level is 300 mV, although signals up to 500 mV can be accommodated. In SSB and DSBSC radio systems it is common to generate the sideband signal at a single frequency, and then frequency translate it to the operating frequency in a heterodyne mixer.

A product detector circuit for the LM-1496 and LM-1596 is shown in Fig. 7-17. In this case the SSB or DSBSC signal is applied to the +IN, while the carrier is applied to the −CARRIER input. The demodulated audio appearing on pin no. 12 will contain a residual of the carrier and RF signal, and that must be stripped off in an *RC* low-pass filter. The DC level on the output is stripped off by the coupling capacitor.

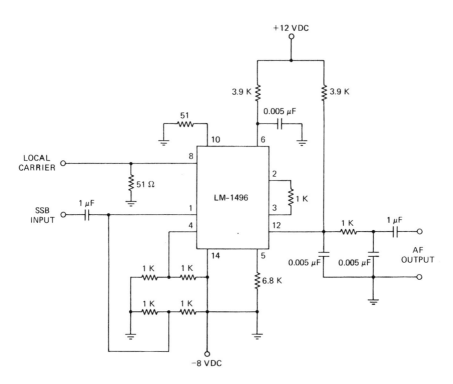

FIGURE 7-17 LM-1496 balanced modulator.

7-5.2 Pulse width modulator circuits

A pulse width modulator (PWM) produces a squarewave output in which the duration of each cycle varies with the modulating signal. Another name for this type of circuit, which is equally descriptive, is *duty-cycle modulator*. The circuit for a basic PWM is shown in Fig. 7-18A. The PWM is nothing more than a voltage comparator used such that the modulating signal (V_m) is applied to one input of the comparator, while the carrier signal (V_c) is applied to the alternate input of the comparator. In the case shown in Fig. 7-18A the modulating signal is applied to the noninverting input, while the carrier is applied to the inverting input of the comparator. This PWM circuit is therefore an inverting pulse width modulator. The carrier signal should be a triangle or sawtooth waveform.

The timing waveforms for the PWM are shown in Fig. 7-18B. The upper plot shows the triangle wave carrier (V_c) and the modulating signal (V_m). In electronic instrumentation applications the varying modulating signal might be an audio carrier, or (more likely) a time-varying DC potential that represents a physical parameter such as temperature or pressure.

According to the normal rules for a voltage comparator: output V_o is LOW whenever $V_c < V_m$, and HIGH whenever $V_c > V_m$. Thus, the output pulse

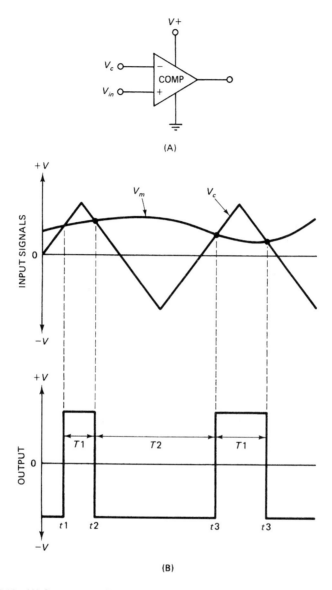

FIGURE 7-18 (A) Comparator; (B) input and output signals.

will snap HIGH on the peaks of the carrier signal, and will remain HIGH a time $t1$ that is inversely proportional to the value of the modulating signal.

The duty cycle of the output pulse produced by Fig. 7-18A (see lower plot in Fig. 7-18B) is given by:

$$D = \frac{T1}{T2} \qquad (7\text{-}38)$$

or, related to the applied voltages:

$$D = \frac{1 - (V_m/V_c)}{2} \qquad (7\text{-}39)$$

EXAMPLE 7-2

A pulse width modulator such as Fig. 7-18A uses a 0 to +6 volt triangle waveform carrier signal. Calculate the duty cycle (D) when a constant potential of +3.5 Vdc is applied to the modulation input.

Solution

$$D = \frac{1 - (V_m/V_c)}{2}$$

$$D = \frac{1 - (3.5 \text{ volts}/6 \text{ volts})}{2}$$

$$D = \frac{1 - 0.583}{2} = 0.21 \qquad \text{(i.e. 21\%)} \qquad \blacksquare$$

In the example above a constant DC voltage was used as the modulating signal. In some instrumentation applications the modulating signal will be a DC level, or slowly varying DC level (as in the case of a temperature signal), or it may be a more dynamic signal. In the latter case Eq. (7-38) can be used to calculate the instantaneous duty factor if the instantaneous values of the carrier signal and modulating signal are used.

7-5.3 FM demodulators

Over the years several methods have been used to demodulate FM and PM signals. Of these, however, only two are in widespread use in IC-based FM/PM receivers: *pulse counting detectors* and *IC quadrature detectors* (ICQD).

The FM/PM receivers typically receives a VHF/UHF RF signal, and then frequency translates it down to an intermediate frequency (IF) of 10.7 MHz (or something similar). A few two-way radio receivers further down-convert the 10.7 MHz IF to a second IF in the 455 kHz range. The latter are called *dual conversion receivers*, while the former are called *single conversion receivers*; both are examples of *superheterodyne receivers*.

The pulse counting FM detector is shown in Fig. 7-19A. An input stage ($A1$) is responsible for squaring the input FM IF sinewave signals. Either a voltage comparator, high-gain limiting amplifiers, or a Schmitt trigger may be used in this function. The output of this stage is a squarewave containing the same frequency or phase variations as the input signal.

The pulses from A1 are used to trigger a dual output, one-shot multivibrator ($OS1$). The duration of $OS1$ output pulse must be much shorter than the minimum width of the input signals. The Q and NOT-Q outputs of

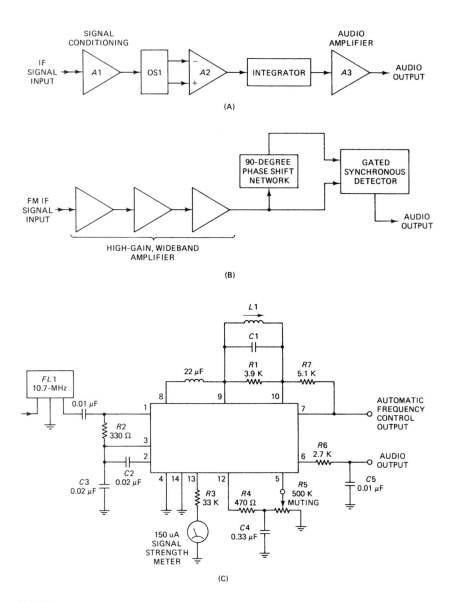

FIGURE 7-19 (A) Coilless FM detector; (B) block diagram of FM quadrature detector; (C) CA-3089 quadrature FM detector.

$OS1$ are applied to a differential amplifier level translator. The output of the differential amplifier is a train of squarewave pulses which have a constant amplitude and constant duration, but vary in repetition rate with changes of input frequency. If these pulses are integrated, then they produce a varying output voltage that is the reconstructed modulating audio.

COMMUNICATIONS APPLICATIONS OF LINEAR IC DEVICES   **237**

The IC quadrature detector is shown in block form in Fig. 7-19B, and as a complete IC circuit in Fig. 7-19C.

Refer first to Fig. 7-19B. The input signal is applied first to a high-gain cascade chain of wideband amplifiers that has the effect of squaring the input sinewave signal. The clipping action is caused by the fact that, for sinewaves, gain is excessive. The clipping serves to remove amplitude peaks on which the majority of noise rides (note: most noise signals amplitude modulate the signal).

The squarewave signal output from the wideband amplifier is split into two components. One component is fed directly to the gated synchronous quadrature detector stage (and is called the 0° signal). The other signal is phase shifted −90° in an external RLC network. The −90° signal is also fed to the detector.

The gated synchronous detector serves as a quadrature phase detector in order to recover the modulating audio.

Figure 7-19C shows the circuit for an IC quadrature detector (ICQD) based on the CA-3089 (also called LM-3089) device. The input side of the ICQD is tuned by a ceramic or crystal bandpass filter ($FL1$) to the 10.7 MHz IF frequency. The filter bandpass is selected to accept the deviation of the FM carrier under modulation.

The phase shift network ($R1/C1/L1$) is tunable so that it can be adjusted to account for tolerances in the IF center frequency. These tolerances are mostly due to variations in filter FL1.

7-6 PHASE LOCKED LOOP (PLL) CIRCUITS

The *phase locked loop* (PLL) circuit was invented in the 1930s as a synchronous AM demodulator. Oddly enough, its first intended use never caught on except in a few overseas shortwave relay receiver sites. Few, if any, other AM detectors are based on the PLL circuit. A host of other applications for the PLL are found, however: touchtone decoding, frequency shift keying (FSK) decoding, FM demodulation, FM/FM telemetry data recovery, FM multiplex stereo decoding (to reconstruct the pilot signal), motor speed control, and transmitter frequency control.

As a transmitter frequency controller the PLL has shown immense popularity from cheap CB transmitters to the most expensive broadcast models. The PLL now dominates the frequency control sections of radio transmitters. The popularity of the PLL in frequency synthesizers is that they allow multichannel fixed and discrete variable frequency control, yet the output frequency has the same stability as the crystal oscillator used as the PLL reference frequency.

Figure 7-20 shows the block diagram for a basic phase locked loop circuit. The main elements of the PLL circuit are a *voltage controlled oscillator* (VCO), *phase detector, reference frequency source,* and a *low-pass filter*

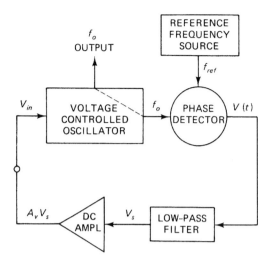

FIGURE 7-20 Simple phase locked loop.

(LPF). A *DC amplifier* may also be used for scaling or level translation of the DC control voltage from the output of the low-pass filter.

The VCO is a special form of variable frequency oscillator in which the output frequency f_o is a function of the input control voltage (V_{in}). In Fig. 7-20 the VCO input voltage is also the output voltage from the DC amplifier.

The reference frequency source is a stable oscillator operating on a fixed frequency. The reference frequency (f_{ref}) in Fig. 7-20 is equal to, or less than, f_o.

The phase detector is a circuit that compares two signals and generates an output that is proportional to the phase difference between them. These circuits are discussed further below. In a PLL circuit the phase detector output ($v(t)$) will be either a pulse train (in digital phase detector) or a DC voltage (analog phase detectors). In both the digital and analog phase detector circuits the output must be processed in a low-pass filter in order to remove residuals of f_o and f_{ref}. In the digital case, the low-pass filter also serves to create the DC control voltage by integrating the pulse train produced by the phase detector.

A modified form of phase detector is shown in Fig. 7-21. This circuit is the more common of the two in transmitter frequency control, signal generator and similar applications where presettable, discrete frequencies are needed. The basic difference between these circuits is the *divide-by-N counter* between the VCO output and the phase detector input port. The reference frequency must be an integer subharmonic of the desired output frequency that will be produced by the VCO: $f_{ref} = f_o/N$.

COMMUNICATIONS APPLICATIONS OF LINEAR IC DEVICES

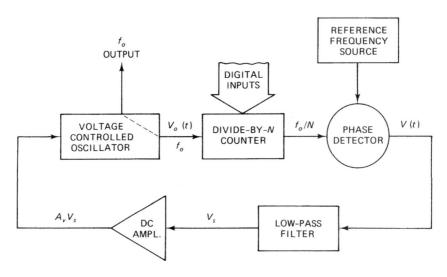

FIGURE 7-21 Phase locked loop frequency synthesizer.

The divide-by-N digital counter can be used to change the frequency of the VCO. If the division ratio of the counter is changed, then the VCO will be forced to the new frequency that maintains the equality $f_{\text{ref}} = f_o/N$.

There are three modes of operation in a PLL: *free-running*, *capture* (also called *search*) and *locked*. In the free-running mode the VCO is not under control, and operates on an essentially random frequency within its range. The PLL is typically in the free-running mode for a brief period after turn/on, but beyond that will be free-running only if a defect is present. In the capture mode the PLL is attempting to lock onto the correct frequency. The VCO frequency tends to converge onto the desired frequency. When the VCO reaches the correct frequency, and remains there, the PLL is said to be in the locked mode.

The reference frequency source controls the output frequency because it is compared with the VCO output in the phase detector. When there is a difference between the VCO output and the reference frequency, a DC control voltage is generated that tends to pull the VCO onto the correct frequency. Thus, the PLL is a form of feedback control system, or 'electronic servomechanism'. The reference frequency sets the minimum step between discrete VCO frequencies.

The stability of the PLL is set by the stability of the reference frequency source. In the most stable systems, such as signal generators or transmitter channel controllers, the reference frequency will be a crystal oscillator that is either temperature compensated or operated inside of a temperature stabilization oven. The output frequency of the crystal oscillator may be divided in a divide-by-N chain of digital counters to produce a low frequency such as 5 kHz, 1 kHz or 100 Hz (which are not easily obtained in crystal oscillators). At least one signal source locks the reference frequency to the extremely accurate and stable 60 kHz WWVB signal broadcast by the US Government from the standard station at Fort Collins, CO.

7-6.1 Phase sensitive detector circuits

The job of the phase sensitive detector (PSD) circuit is to generate an output that is proportional to the difference in-phase between two input frequencies. Several different forms of PSD circuit are typically used in PLL circuits, but in this brief discussion only the digital form will be covered. Two basic phase sensitive detector are used in most digitally controlled IC PLL circuits: the *exclusive-OR gate detector* (Fig. 7-22A) and the *R-S Flip-Flop edge detector* (Fig. 7-24A). Although both of these circuits are somewhat more sensitive to harmonics, noise and reference source jitter, they are sufficiently easy to implement (and sufficiently useful) to cause them to see widespread application.

The exclusive-OR (XOR) gate, shown in Fig. 7-22A, is designed to output a HIGH only when the two inputs are different from each other (one is HIGH and the other is LOW). When the two inputs are at the same level, either both LOW or both HIGH, then the output of the XOR gate is LOW. Figure 7-22B shows the truth table for the XOR gate. In the digital PLL circuit, the inputs of the XOR gate are excited with the squarewave f_o and f_{ref} signals.

Figure 7-23 shows several cases for the relationship between the reference and output frequencies as applied to inputs A and B of the XOR gate. Despite the fact that these signals are squarewaves, the convention of using radian notation is followed. The start of the cycle is designated with zero or 2π, while the point halfway through the cycle is designated π. In Fig. 7-23A the two input signals are completely out of phase with each other ($\phi = \pi$). This condition forces one of the inputs HIGH all of the time, but neither is HIGH at the same time. In other words, the two inputs always see a different level at the two inputs. If one is HIGH, then the other is LOW, and vice versa.

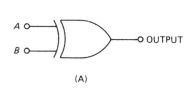

FIGURE 7-22 Exclusive-OR (XOR) gate.

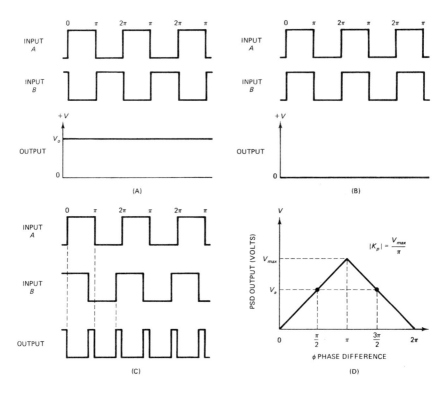

FIGURE 7-23 XOR-gate phase detector outputs under different phase conditions.

Thus, the output of the XOR gate PSD is always HIGH for this condition. It is essentially a DC level.

The opposite condition is shown in Fig. 7-23B. Here the two signals are identical to each other, i.e. they are in-phase with one another. In this condition, the two inputs of the XOR gate PSD are always the same, either HIGH or LOW. Thus, the output of the PSD is perpetually at zero. It is established that the output of the XOR gate PSD will be zero for in-phase signals, and maximum for 180° (π) out-of-phase signals. The condition for phase differences less than π is shown in Fig. 7-23C. The output of the XOR gate PSD is a train of short duration pulses with a width that is proportional to the difference in-phase between the two input signals.

A plot of the output of the XOR gate PSD is shown in Fig. 7-23D. Note that the output voltage is maximum at $\phi = 0$, and zero at both $\phi = 0$ and $\phi = 2\pi$. The slope between the maxima and minima is the *conversion gain* (K_p). A problem with this type of detector can easily be seen in this graph. From a center phase difference of π (180°), where the voltage output is maximum, the voltage drops off linearly to zero. But the graph is symmetrical about 180°. The same voltage (V_a) therefore represents two different phase

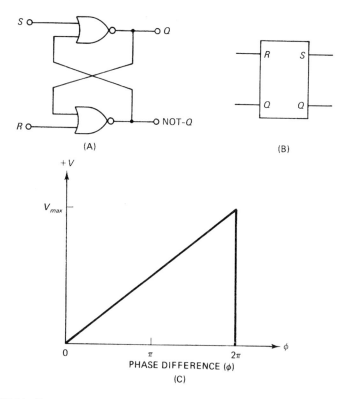

FIGURE 7-24 Reset-set flip-flop: (A) circuit; (B) schematic symbol; (C) output function of phase detector.

differences, in this example $\pi/2$ and $3\pi/2$. The edge detector PSD circuit eliminates this problem at the cost of slightly greater complexity.

The *reset-set flip-flop* (R-S FF) circuit shown in Fig. 7-24 can be used for phase detection. The R-S FF can be constructed of either NAND or NOR gates, but in PSD circuits it is the NOR-logic R-S FF that is used. The two NOR gates in Fig. 7-24A are cross-coupled so that the output of one becomes the input of the other. The schematic symbol shown in Fig. 7-24B is sometimes used to represent the R-S FF, especially in the cases where the gate is found in a TTL or CMOS digital IC.

The rules of operation for the NOR-logic R-S FF are simple. When a brief positive-going pulse is applied to the SET input, the Q-output is forced to HIGH and the NOT-Q output goes LOW (Q and NOT-Q are always opposite each other because they are complementary). The Q-output will remain HIGH until a second positive-going pulse is applied to the RESET input.

When used as a phase sensitive detector the two inputs are connected to the two difference frequency sources. The SET input is connected to the reference frequency (f_{ref}) and the RESET input to the VCO frequency (f_o).

The effect of using the NOR R-S FF is shown in Fig. 7-24C. Note that the voltage rises linearly from zero to V_{max} over the range 0 to 2π, which represents the entire potential 360° of the phase difference. Above the 2π point, however, the voltage drops to zero.

7-6.2 Phase locked loop IC devices

Although the phase locked loop (PLL) circuit was invented in the 1930s, it required the modern integrated electronics technology to make the PLL practical enough for a large range of applications. In this section we will examine several of the more popular PLL IC devices in order to give the student some idea of what's available.

MC-4044/4344 PLL Chip. The MC-4044 PLL device is shown in block diagram form in Fig. 7-25A. The 4044 device is used in a wide variety of applications including transmitter frequency control. In most cases, the 4044 is used in conjunction with other devices of the same family in order to extend its range of applications. The internal circuitry of the 4044 includes a pair of identical PSD circuits, a charge pump and a DC amplifier. External circuitry links these elements together to form the PLL circuit.

The phase sensitive detectors are the digital type, and are fed in parallel from the same pair of inputs. The reason why two PSDs are used is to permit quadrature phase detection. If an ordinary PSD circuit is needed, then only one phase detector is used. But if the quadrature form is needed, then both internal phase detectors are used. The advantage of the quadrature form of PSD is that it is less sensitive to noise, and rejects certain harmonics of the input signals.

The charge pump is used to drive an external capacitor that is used as part of the low-pass filter network. The voltage across the charge pump capacitor is used as the DC loop control voltage. The DC amplifier is used to scale the DC control voltage as needed.

The 4044 has a transfer function as shown in Fig. 7-25B. The conversion gain (K_p) is on the order of 120 mV/radian when ordinary supply voltages are used. Note that the phase detectors are linear over the range $-360°$ to $+360°$, which represents a 720° capture range.

MC-145152 PLL IC. Another PLL chip is diagrammed in Fig. 7-26: the MC-145152. This IC contains three divide-by-integer counters: counter R is programmable to divide by 8, 64, 128, 256, 512, 1024 or 1160; counter N is programmable to any value between 3 and 1023; counter A is programmable to any value between 0 and 63. The R counter is used to process the reference signal (f_{ref}), while the others process the VCO output signal. The chip also contains a phase sensitive detector with differential outputs, and a control logic section that governs the 'housekeeping' of the device.

Figure 7-27 shows the block diagram of a PLL circuit developed by RF Prototype Systems of San Diego, CA. This PLL can be designed to operate

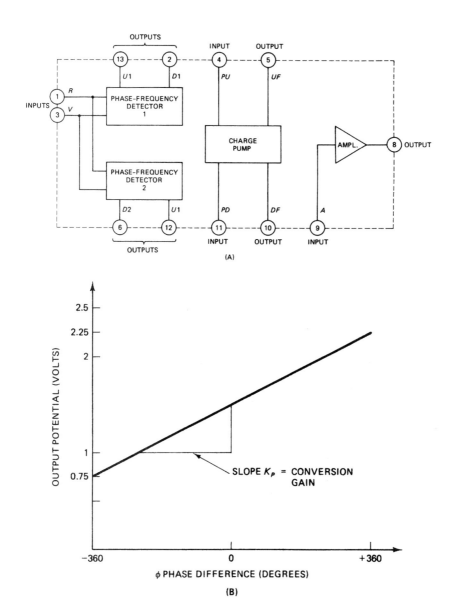

FIGURE 7-25 (A) MC-4044 PLL chip; (B) output function.

at any frequency between 35 MHz and 920 MHz. The specific frequency range, and spacing between channels, depends upon the design of the circuit and the particular voltage controlled oscillator (VCO) used.

The division ratios of the R, N and A dividers inside the MC-145152 are set by external switches. The setting of these switches determines the

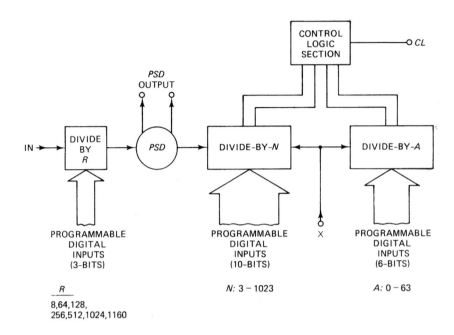

FIGURE 7-26 MC-145152 chip.

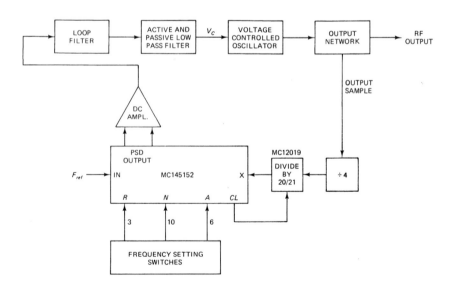

FIGURE 7-27 MC-145152 circuit.

operating frequency of the overall circuit. In practical circuits these switches may be any of several forms: DIP switches, front panel switches, a numerical ASCII touchpad, or a digital register that is loaded from a computer or other digital source.

The differential outputs of the PSD are used to drive a DC differential amplifier, which in turn drives the loop filter and a cascade series of active

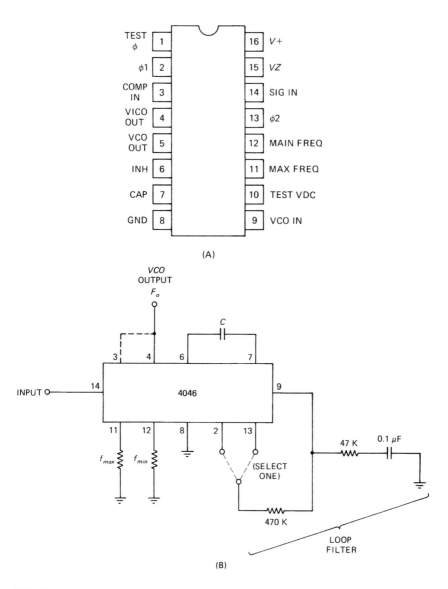

FIGURE 7-28 4046 PLL chip: (A) pinouts and (B) circuit.

and passive low-pass filters. The output of these filters is the DC control voltage (V_c) that actually controls the VCO output frequency. The output network isolates and impedance matches the VCO and the RF output. The output network also samples the RF output signal to form the signal that is used to compare with the reference frequency. This output sample may be frequency divided (as shown) in two or more stages, or it may be applied directly to the input of the MC-145152 inputs.

CMOS 4046 PLL IC. The 4046 CMOS device (Fig. 7-28A) is a phase locked loop. Do not confuse the CMOS 4000-series device with the MC-4044 device considered earlier. In the CMOS 4000-series of devices the 4044 is a quad NAND-logic R-S flip-flop, not a PLL. The MC-4044, which is a PLL, uses a different technology and is not part of the CMOS series of chips.

The 4046 contains two phase sensitive detectors. $\phi 1$ is an XOR-gate based PSD, and is somewhat sensitive to harmonics in the input signals. It is, however, also simple. Phase detector $\phi 2$ uses a different form of logic, and is capable of a frequency range up to 2000:1. It will also accept any duty factor, while $\phi 1$ is limited.

The PSD (whichever is selected) is routed to the VCO input (pin no. 9) through an *RC* loop filter (see Fig. 7-28B). The operating frequency of the 4046 PLL device is set by a combination of the input voltage applied to pin no. 14, and the parameter setting components applied to pins 6, 7, 11, and 12. A capacitor (≥ 50 pF) is connected between pins 6 and 7. The maximum frequency resistor is connected to pin no. 11, and must be between 10 kohms and 1 megohm. The minimum frequency resistor is connected to pin no. 12, and must be larger than the maximum frequency resistor. The minimum resistor must be >10 kohms, but there is no upper limit.

NE/SE-565 PLL IC. One of the earliest IC PLL devices was the NE/SE-565 by Signetics. The 565 is part of a family of PLL devices that includes other PLL ICs and a VCO IC (NE/SE-566). The NE-565 operates over the commercial temperature range (0 to $+70°C$), while the SE-565 operates over the harsh commercial and military temperature range ($-40°C$ to $+125°C$). The operating voltage can be ± 6 Vdc to ± 12 Vdc. The operating frequency range is 0.001 Hz to 550 kHz. The 565 (Fig. 7-29) contains a single phase detector, a DC amplifier, and a voltage controlled oscillator. The loop filter consists of a pair of components: resistor R1 and capacitor $C1$.

7-6.3 Loop filter circuits

The loop filter in a PLL circuit is a low-pass filter that serves three basic functions. First, it produces the DC feedback control voltage that is used to set the VCO frequency. Second, it sets the responsiveness of the PLL. A very heavily damped loop filter will force the PLL to be sluggish and hard to change (which might be an advantage if there is a large amount of noise present). Third, the loop filter also removes any residual components of the

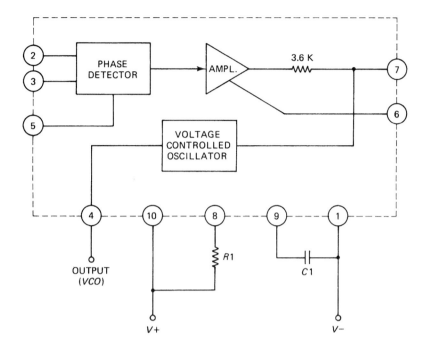

FIGURE 7-29 NE-565 chip.

VCO and reference frequency that are present in the PSD output. Several different types of circuit are used for the loop filter (Fig. 7-30).

Figures 7-30A through 7-30C show several passive *RC* network low-pass filter that are often used as loop filters in PLL circuits. In Figs 7-30D and 7-30E are two active loop filters based on operational amplifiers. The −3 dB cutoff in the frequency response of these first-order low-pass filters is set by:

$$f_{-3 \text{ dB}} = \frac{1}{2\pi R1C1} \tag{7-40}$$

These filter circuits have a relatively poor slope in the frequency response beyond the −3 dB point, on the order of −6 dB/octave. Second-order and higher active filters may also be used for the loop filter application (see Chapter 8).

7-6.4 A PLL application: sub-audio frequency meter

Sub-audio frequencies are those that are below the range of human hearing, i.e. $f < 20$ Hz. In practical circuits the range of sub-audio frequencies may be on the order 0.0001 Hz to 20 Hz. These frequencies are notoriously difficult to measure in regular digital frequency counters because the gate times

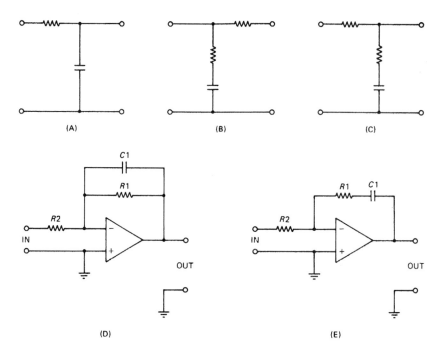

FIGURE 7-30 PLL loop filter circuits.

are too long. In some models, short gate times (1 second) were used, and then the result extrapolated to the actual frequency based on the count during the gate time. This method produces relatively large errors. A better (but more complex) method used in some models is to measure the *period* of the input signal, and then take its reciprocal in an arithmetic circuit. This method was once used extensively in very low frequency counters. A final method, which has certain advantages, is to multiply the input frequency $\times 100$ or $\times 1000$ in a phase locked loop.

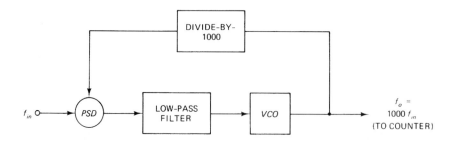

FIGURE 7-31 Sub-audio PLL frequency meter.

Figure 7-31 shows the block diagram of a PLL-based sub-audio frequency meter. The VCO operates over a range that is either 100 or 1000 times the input frequency range. The VCO output is divided in a divide-by-N (where $N = 100$ or 1000) counter before being applied to one input port of a phase sensitive detector (PSD). The sub-audio input frequency is applied to the alternate input port of the PSD. If $N = 1000$, then a 1 Hz signal will read '1000 Hz' on the output counter.

7-7 SUMMARY

1. Current loop circuits can be used in communications applications. The classical teletypewriter uses a 20 mA current loop (obsolete machines used 60 mA), while some instrumentation applications use a 4-to-20 mA current loop for communications.
2. The 4-to-20 mA current loop offers several advantages over other systems: easy daisy chaining of instruments, non-ambiguous zero, and relative freedom from noise impulses.
3. Both 20 mA and 4-to-20 mA current loops can use voltage to current converter circuits (e.g. Howland current pumps, etc.) on the transmit end, and optoisolators on the receive end. This flexibility makes the circuits relatively simple.
4. The EIA RS-232C standard for serial communications ports uses voltage levels to represent '1' and '0' binary numbers. The '1' level is represented in RS-232C by a voltage between -5 and -15 Vdc, while the '0' level is represented by a voltage of $+5$ to $+15$ Vdc. In each case, potentials between 0 and |5 volts| are regarded as undefined.
5. Modulation is the process of altering a characteristic of one signal, called the *carrier*, in proportion to another signal, called the *modulating signal*, for the purpose of transmitting information or data. Modulation may vary either the frequency, phase or amplitude of the carrier signal.
6. Demodulation is the process of decoding the modulated carrier signal in order to recover the information represented by the original modulating signal.
7. Amplitude modulation can be performed by analog multiplier circuits because AM is basically a multiplication process. The output signal is the product of carrier and modulating frequencies.
8. Two different forms of angular modulation are found: *phase modulation* and *frequency modulation*. Several forms of amplitude modulation are found: *amplitude modulation, double sideband suppressed carrier* (DSBSC), and *single sideband suppressed carrier* (SSBSC or SSB). The SSBSC form is sub-divided into *upper sideband* (USB) and *lower sideband* (LSB). Pulse modulation forms include *pulse position modulation* (PPM), *pulse width modulation* (PWM) and *pulse amplitude modulation* (PAM).
9. Phase locked loop (PLL) circuits create an output frequency from a voltage controlled oscillator (VCO) by comparing the VCO output to a stable reference source. The principal elements of a PLL circuit include the VCO, a phase sensitive detector, a low-pass loop filter, and (in some cases) a DC scaling amplifier.

7-8 RECAPITULATION

Now return to the objectives and Pre-quiz questions at the beginning of the chapter and see how well you can answer them. If you cannot answer certain questions, place a check mark to each and review the appropriate parts of the text. Next, try to answer the questions and work the problems below, using the same procedure.

7-9 STUDENT EXERCISES

1. Design, build and test a 20 mA current loop communications system in which MARK is 16 mA to 20 mA, and SPACE is 0 to 2 mA.
2. Design, build and test a 4-to-20 mA current loop instrumentation communications circuit in which a 2 Vdc signal is represented by 4 mA and 10 Vdc is represented by 20 mA. Optional: include a 1-bit test circuit that detects the off/fault state in which $I \ll 4$ mA when $V \geq 2$ Vdc.
3. Design, build and test a sub-audio frequency meter based on the 4046 CMOS PLL IC.
4. Design, build and test a pulse width modulator (PWM) based on (a) NOR-logic R-S FF, and (b) XOR-gate.
5. Design, build and test an FM demodulator based on the pulse counting technique that will decode FM signals between 500 Hz and 5000 Hz.

7-10 QUESTIONS AND PROBLEMS

1. In a 20 mA current loop communications system, such as a teletypewriter, the MARK (logical 1) is represented by a current of _____ to _____ mA, while a SPACE (logical 0) is represented by a current of _____ to _____ mA.
2. Some process instrumentation communications systems use a _____ to _____ mA current loop.
3. In the question above, a current of 0.2 mA may indicate a _____.
4. A floating load current source (Fig. 7-5) has a 10 kohm load resistor, R_L. Assuming that $R_L \gg R1$, calculate the current when the output voltage is 10 Vdc and the input voltage is 2 Vdc.
5. A Howland current pump (Fig. 7-6) has two input voltages $V1 = +6$ Vdc and $V2 = +1$ Vdc. Assuming that $R = 10$ kohms, calculate the output current in R_L.
6. A Howland current pump is used to make a 4-to-20 mA current loop transmitter. What value of bias voltage $V2$ will force I_L to be 4 mA when input voltage $V1 = +2$ Vdc. Assume $R = 1000$ ohms.
7. List the voltage levels that represent logical 1 and logical 0 in the standard EIA RS-232C standard for serial communications ports.
8. Define modulation in your own words.

9. List several characteristics of a carrier signal that can be varied in a modulation process.
10. List the forms of angular modulation.
11. List the forms of amplitude modulation.
12. List the forms of pulse modulation.
13. What type of sinusoidal modulation produces a sum and difference output frequency, but no carrier frequency?
14. Write the equation for *modulation index* in AM systems.
15. An ordinary amplitude modulation system has a 500 kHz sinusoidal carrier, and is modulated by a 10 kHz sinusoidal modulating signal. Assuming that no frequency selective output filtering is used, what frequencies are present in the output signal?
16. A voltage _____ IC can be used as a pulse width modulator.
17. List two forms of FM demodulator commonly used in IC circuits.
18. List the elements of a phased locked loop (PLL), including any optional elements.
19. Draw the block diagram of a PLL in which the output frequency can be set by applying a digital control signal.
20. List two forms of digital phase sensitive detector circuit. Draw the transfer function V_o versus phase for both types.
21. What is the purpose of $D1$ in Fig. 7-1? What PIV rating should it have?
22. Sketch the circuit of an op-amp based TTL-to-RS-232 converter.
23. Sketch the circuit of an RS-232-to-TTL converter.
24. Sketch the frequency spectrum of the following signals: (a) unmodulated, unkeyed CW, (b) AM, and (c) DSBSC.
25. Sketch the waveform of a modulated signal consisting of a squarewave carrier and a triangle wave modulating signal.

CHAPTER 8

Analog multipliers and dividers

OBJECTIVES

1. Learn the applications of the analog multiplier/divider.
2. Understand how multipliers and dividers work.
3. Learn other multiplier/divider applications, devices and circuits.

8-1 PRE-QUIZ

These questions test your prior knowledge of the material in this chapter. Try answering them before you read the chapter. Look for the answers (especially those you answered incorrectly) as you read the text. After you have finished studying the chapter try answering these questions again, and those at the end of the chapter (see Section 8-23).

1. List three different forms of analog multiplier circuit.
2. An analog multiplier has a scaling factor of 1/20. Calculate the output voltage when $V_x = +1.5$ volts and $V_y = +2$ volts.
3. Draw the circuit for using an AD-533 multiplier square rooter.
4. Prove that an analog multiplier connected as a squarer will operate as a frequency doubler when a sinewave signal is applied.

8-2 ANALOG MULTIPLIER AND DIVIDER IC DEVICES

Analog multiplier and divider circuits are available in both monolithic integrated circuit and hybrid circuit forms. Analog multipliers produce an output voltage V_o that is the product of two input voltages, V_x and V_y. The general form of the multiplier transfer function is:

$$V_o = KV_xV_y \qquad (8\text{-}1)$$

where:
  V_o is the output potential in volts
  V_x is the potential (in volts) applied to the X input
  V_y is the potential (in volts) applied to the Y input
  K is a constant (usually 1/10)

If the proportionality constant K is 1/10, then Eq. (8-1) becomes:

$$V_o = \frac{V_xV_y}{10} \qquad (8\text{-}2)$$

There are several different basic designs for analog multiplier circuits. In Chapter 2 the logarithmic amplifier was discussed and its use as a multiplier will be reviewed briefly here. When the outputs of two logarithmic amplifiers are first summed together, and then applied to an antilog amplifier, the output of the antilog amplifier is proportional (via K) to the product of the two input voltages. Transconductance amplifiers can also be used to make an analog multiplier. There are examples of multipliers based on the operational transconductance amplifier (OTA) IC device. There is also a type called the *transconductance cell* analog multiplier. Other varieties of multiplier circuit will also be examined in this chapter.

8-3 ANALOG MULTIPLIER CIRCUIT SYMBOLS

Figure 8-1 shows typical symbols used to represent analog multiplier and divider circuits in schematic diagrams. Although there are standards for

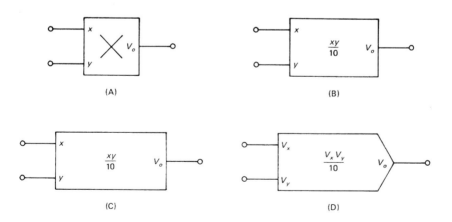

FIGURE 8-1 Analog multiplier circuit symbols.

circuit symbols, the multiplier is one type of device in which corporate, commercial (e.g. IEEE) and military standards all seem to be used simultaneously. The symbols shown in Fig. 8-1 are several of those commonly found, and the reader should be aware that other symbols (as well as variations of these) may well be used in actual practice.

8-4 ONE- TWO- AND FOUR-QUADRANT OPERATION

Analog multipliers and dividers are classified according to the number of quadrants in which they will operate. The quadrants are the four quadrants of the Cartesian coordinate system. Figure 8-2A illustrates one-quadrant operation. In this type of system both input voltages must be positive ($V_x \geq 0$), ($V_y \geq 0$). The only possible output voltage polarity is positive. At least one commercial hybrid multiplier operates in one quadrant, but with both input voltages negative. Again, the only permissible output voltage was positive. That type of operation, is a rarity, however. The least complex multipliers based on logarithmic amplifiers are normally one-quadrant devices.

A second form of multiplier is the two-quadrant form (Fig. 8-2B). These circuits operate in such a manner that allows the output voltage to be either positive or negative, but there are constraints on the allowable input voltage polarities. One input voltage will be limited to positive values only, while the other can be either positive or negative.

Four-quadrant operation (Fig. 8-3) is the most flexible because it allows operation with any combination of input signal polarity. The output signal can be either positive or negative, as can either (or both) input signal voltages. Figure 8-4 shows the relationship between input and output polarities,

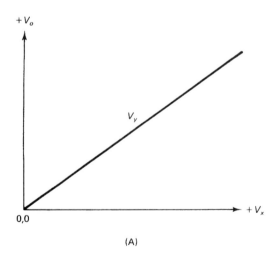

FIGURE 8-2 Analog multiplier transfer functions: (A) single quadrant; (B) two quadrant.

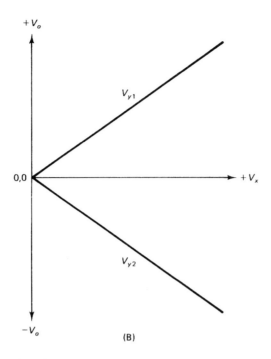

FIGURE 8-2 (continued)

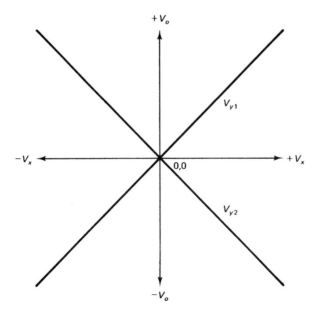

FIGURE 8-3 Analog multiplier transfer function for four-quadrant operation.

ANALOG MULTIPLIERS AND DIVIDERS **257**

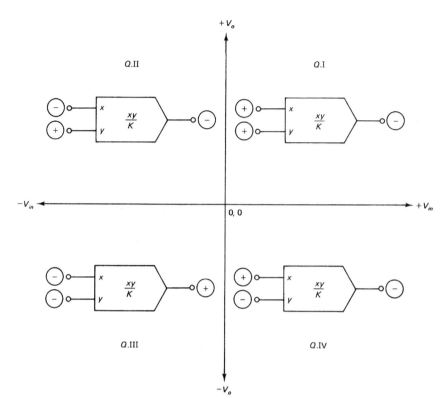

FIGURE 8-4 Input and output conditions for four-quadrant operation.

relating same to the quadrant of operation. These limits are summarized as follows:

$$QI : V_x \geq 0, \ V_y \geq 0, \ 0 < V_o < +V.$$
$$QII : V_x \leq 0, \ V_y \geq 0, \ -V < V_o < 0.$$
$$QIII : V_x \leq 0, \ V_y <= 0, \ 0 < V_o < +V.$$
$$QIV : V_x \geq 0, \ V_y \leq 0, \ -V < V_o < 0.$$

In most cases, the voltage ranges of the two inputs are symmetrical with each other. For example, a typical range is $V_{x(max)} = V_{y(max)} = \pm 10$ volts.

Another category of device is the multiplier/divider shown in Fig. 8-5. These devices have a transfer function of the form:

$$V_o = \frac{V_x V_y}{V_z} \quad (8\text{-}3)$$

In the multiplier mode, the X and Y inputs are used, and the scaling factor K is set by applying a voltage to the Z input. In the division mode, either

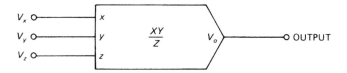

FIGURE 8-5   XY/Z multiplier/divider.

the X and Z, or Y and Z inputs are used for the signal voltages, while the scaling factor is set by applying a fixed voltage to the remaining input.

8-5   DIFFERENTIAL INPUT MULTIPLIERS AND DIVIDERS

Most of the multiplier and divider circuits presented in this chapter are single-ended devices. That is, the input signals are measured between the input terminal and ground, and there is but one input line each for V_x and V_y. A differential input multiplier and/or divider circuit is shown in Fig. 8-6. In this circuit the differential input voltages V_{dx} and V_{dy} are defined as:

$$V_{dx} = X1 - X2 \tag{8-4}$$

$$V_{dy} = Y1 - Y2 \tag{8-5}$$

The multiplier transfer function is:

$$V_o = K V_{dx} V_{dy} \tag{8-6}$$

or

$$V_o = K(X1 - X2)(Y1 - Y2) \tag{8-7}$$

In most cases, $K = 1/10$, so these expressions become:

$$V_o = \frac{V_{dx} V_{dy}}{10} \tag{8-8}$$

$$V_o = \frac{(X1 - X2)(Y1 - Y2)}{10} \tag{8-9}$$

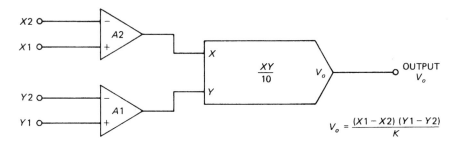

FIGURE 8-6   Differential multiplier/divider.

The differential input multiplier is particularly useful in at least two different types of situation. First, where a Wheatstone bridge or other balanced output signal source supplies one or both multiplier input signals. Second, in cases where the signal processing requirements are such that an input to the multiplier is the difference between source signals. For example, if a fluid pressure transducer measures a pressure in which a variation rides on top of a certain large offset minima pressure (e.g. the human blood pressure, in which the varying pressure wave rides atop the diastolic static pressure), then the input signal will be a slowly varying DC wave on top of a large DC static offset potential. If the DC offset can be subtracted prior to being applied to the multiplier, then the overall dynamic range of the system is improved.

Either input of Fig. 8-6 can be made single-ended by the simple expedient of grounding the unwanted input.

8-6 TYPES OF MULTIPLIER CIRCUIT

In this section we will examine several popular approaches to multiplier design. Some of them are currently used in state-of-the-art analog multipliers, while others are now obsolete. The reason for including the obsolete circuits is to demonstrate the process of multiplication and the range of possibilities. Besides, it is generally an error to disdain older methods because the same fundamentals are often resurrected when newer methods can make their implementation better. For example, the incandescent lamp multiplier circuit below is now obsolete, but the JFET manifestation of the same concept is currently used. Yet is it nonetheless profitable to examine the lamp version because it is inherently easier to understand.

8-7 MATCHED RESISTOR/LAMP MULTIPLIER DIVIDERS

A simple and effective multiplier and divider circuit can be constructed from a pair of operational amplifiers and a special gain setting block (see Fig. 8-7A) consisting of a pair of matched photoresistors ($R1$ and $R2$), both of which are illuminated by the same incandescent lamp ($IL1$). Because $R1$ and $R2$ are matched for any given light level, it is true that $R1 = R2 = R$. Because $R1$ is in the feedback loop of amplifier $A2$, nonlinearities in the voltage versus brightness ratio of $IL1$ are 'servoed-out'.

In past chapters we have used an analysis method based on Ohm's law and Kirchhoff's current law (KCL). That practice will be continued here. In Fig. 8-7A, the input bias current of each amplifier (I_{b1} and I_{b2}) are each zero. In addition, the noninverting inputs of both amplifiers are grounded, so the two summing junctions (A and B) are at zero volts (ground) potential. By Ohm's law, and considering that $R1 = R2 = R$:

$$I1 = \frac{V_z}{R} \tag{8-10}$$

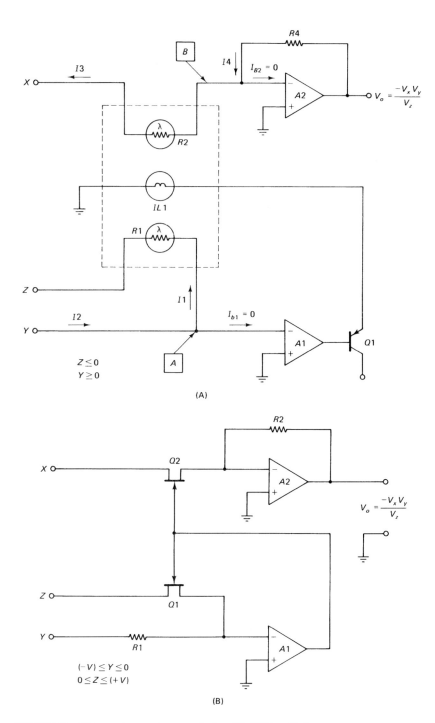

FIGURE 8-7 (A) Incandescent lamp multiplier; (B) JFET multiplier.

and by KCL:

$$I2 = \frac{V_y}{R3} \tag{8-11}$$

$$I1 = -I2 \tag{8-12}$$

Substituting Eqs (8-10) and (8-11) into Eq. (8-12):

$$\frac{V_y}{R3} = \frac{-V_z}{R} \tag{8-13}$$

and, rearranging terms:

$$R = \frac{-V_z R3}{V_y} \tag{8-14}$$

Current $I3$ is, by Ohm's law:

$$I3 = \frac{V_x}{R} \tag{8-15}$$

so, by substituting Eq. (8-14) into Eq. (8-15):

$$I3 = \frac{V_x}{\left[\dfrac{-V_z R3}{V_y}\right]} \tag{8-16}$$

and, then rearranging terms:

$$I3 = \frac{V_x V_y}{V_z R3} \tag{8-17}$$

Because $I_{b1} = 0$, and because point B is at ground potential, by KCL it can be stated:

$$I3 = -I4 \tag{8-18}$$

From Ohm's law:

$$I4 = \frac{V_o}{R4} \tag{8-19}$$

By substituting Eqs (8-17) and (8-19) into Eq. (8-18):

$$-\frac{V_o}{R4} = -\frac{V_x V_y}{V_z R3} \tag{8-20}$$

or, when the negative signs are dropped and the terms are rearranged:

$$V_o = \frac{V_x V_y R4}{V_z R3} \tag{8-21}$$

Accounting for the fact that V_z is restricted to negative values:

$$V_o = -\frac{V_x V_y R4}{V_z R3} \quad (8\text{-}22)$$

Placing Eq. (8-22) into the standard form:

$$V_o = -\frac{K V_x V_y}{V_z} \quad (8\text{-}23)$$

in which $K = R4/R3$.

The circuit of Fig. 8-7A is somewhat impractical today, but it was once implemented in discrete form in analog circuitry. Figure 8-7B shows a modern version in which the lamp and resistors have been replaced with junction field effect transistors (JFET). The circuit of Fig. 8-7B works on the basis of the channel resistance of the JFETs. At gate potentials below the pinch-off voltage, the JFET drain–source resistance is a function of the gate voltage. Thus, $Q1$ and $Q2$ operate as voltage controlled resistors.

8-8 LOGARITHMIC AMPLIFIER-BASED MULTIPLIERS

Many popular forms of analog multiplier–divider circuit are based on the properties of the logarithmic and anti-log amplifiers (see Chapter 2). A logarithmic amplifier has a transfer function of the form either:

$$V_o = k \ln(V_{in}) \quad (8\text{-}24)$$

or

$$V_o = k \log(V_{in}) \quad (8\text{-}25)$$

One of the properties of logarithms is that their use converts multiplication operations into addition, and division operations are converted into subtraction. Therefore:

$$XY = \log X + \log Y \quad (8\text{-}26)$$

and

$$\frac{X}{Y} = \log X - \log Y \quad (8\text{-}27)$$

Equations (8-26) and (8-27) give us the basis for designing an analog multiplier/divider circuit. Figure 8-8 shows the block diagram of such a circuit. The X input (V_x) is applied to the input of LOGAMP A to produce signal $A = \log(V_x)$. Similarly, the Y input (V_y) is applied to the input of LOGAMP B to produce a signal $B = \log(V_y)$. These signals are each applied to a summer amplifier (in the multiplication case) or a difference amplifier (in the division case). The output of the summer circuit is $C = A + B$, or $C = \log(V_x) + \log(V_y)$. By passing this signal through an antilog amplifier

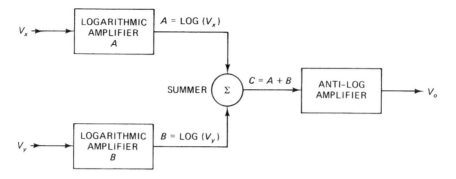

FIGURE 8-8 Log-antilog multiplier.

it is possible to construct the product of X and Y:

$$V_o = \log^{-1}(C) \qquad (8\text{-}28)$$

$$V_o = \log^{-1}[\log(V_x) + \log(V_y)] \qquad (8\text{-}29)$$

$$V_o = V_x V_y \qquad (8\text{-}30)$$

The logarithmic and antilog amplifier were discussed in Chapter 2. That discussion revealed that an uncompensated LOG/ANTILOG amplifier exhibits a strong temperature dependence. In order to prevent error, temperature compensation must therefore be incorporated into the multiplier circuit.

8-9 QUARTER SQUARE MULTIPLIERS

Figure 8-9 shows the block diagram of a quarter square multiplier. Consider the following expression:

$$V_o = \frac{(X+Y)^2 - (X-Y)^2}{4} \qquad (8\text{-}31)$$

Expanding this polynomial results in:

$$V_o = \frac{(X^2 - X^2) + (Y^2 - Y^2) + 2XY + 2XY}{4} \qquad (8\text{-}32)$$

$$V_o = \frac{2XY + 2XY}{4} \qquad (8\text{-}33)$$

$$V_o = XY \qquad (8\text{-}34)$$

Figure 8-9 shows the block diagram of a circuit that will implement Eq. (8-31). There are two chains of circuits in this multiplier. Each chain contains a two-port input amplifier: $A1$ is a summation amplifier and produces

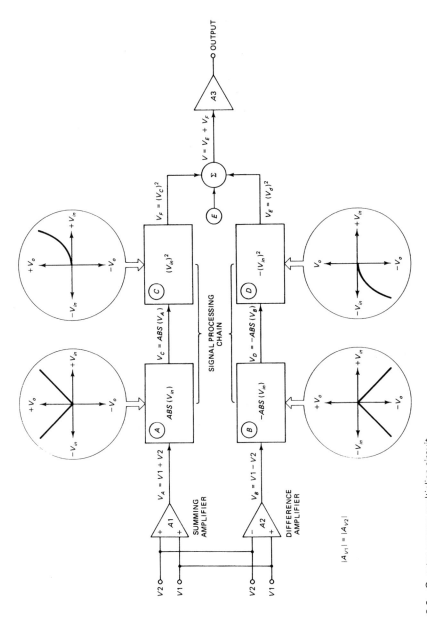

FIGURE 8-9 Quarter square multiplier circuit.

ANALOG MULTIPLIERS AND DIVIDERS

an output voltage $V_a = (V1 + V2)$; A2 is a difference amplifier and produces an output $V_b = (V1 - V2)$.

Each signal processing chain also contains an *absolute value amplifier* and a *squaring circuit*. The absolute value circuit is also called a *fullwave precise rectifier*. The only difference between these two chains is that A is noninverting, while B is inverting. Thus:

$$V_c = \text{abs}(V_a) \tag{8-35}$$

$$V_o = \text{abs}(V1 + V2) \tag{8-36}$$

and

$$V_D = -\text{abs}(V_B) \tag{8-37}$$

$$V_D = -\text{abs}(V1 - V2) \tag{8-38}$$

The squarers (C and D) produce an output that is the square of the input voltage ($V_o = (V_{in})^2$). In analog multipliers the squarers are *diode breakpoint generators*. Ordinary PN junction diodes have a 'square law' region in their operating characteristic. Using as few as ten PN junction diodes, each biased to slightly different points, produces a breakpoint generator with a linearity approaching ±0.1% of full scale. The outputs of the squarers are:

$$V_E = V_D^2 \tag{8-39}$$

$$V_F = V_C^2 \tag{8-40}$$

Combining these voltage in a summer produces:

$$V_o = V_C^2 + V_D^2 \tag{8-41}$$

Relating Eqs. (8-41) to (8-31) is a good exercise for the student.

8-10 TRANSCONDUCTANCE MULTIPLIERS

The most commonly used form of IC analog multiplier is the transconductance type. The word 'transconductance' indicates that an output current (I_o) is controlled by an input voltage (V_{in}):

$$G_m = \frac{I_o}{V_{in}} \tag{8-42}$$

In older texts the unit of conductance was the descriptive term *mho* (ohm spelled backwards), but today the *siemen* is used for conductance (note: it is only a name change, for 1 siemen = 1 mho).

Figure 8-10 shows the circuit for a simple op-amp based transconductance multiplier. The basis for the circuit is a dual NPN transistor such as the LM-114 device. It is important that the two transistors be part of the same substrate

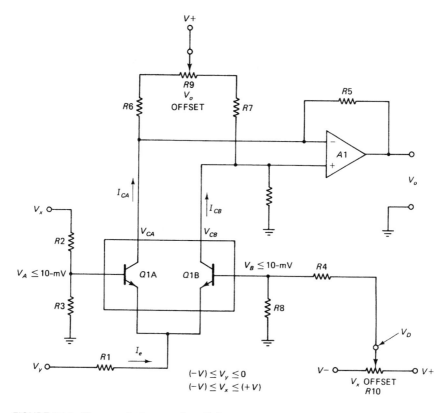

FIGURE 8-10 Transconductance cell multiplier.

in order to maintain thermal tracking between the two devices. For very low signal levels (10 mV or so), the following relationship is true:

$$I_{ca} = kI_e V_x \qquad (8\text{-}43)$$

where:
 I_{ca} is the collector current of $Q1A$
 I_e is the emitter current
 V_x is the voltage applied to the base-emitter junction of $Q1A$

$$k = q/2KT$$

q is the electronic charge 1.6×10^{-19} coulombs
K is Boltzmann's constant $(1.38 \times 10^{-23}$ J/K)
T is the temperature in kelvin (K)

Output voltage V_o is:

$$V_o = I_{ca} R5 \qquad (8\text{-}44)$$

ANALOG MULTIPLIERS AND DIVIDERS

The emitter current is supplied from input voltage V_y, so:

$$I_e = \frac{V_y}{R1} \qquad (8\text{-}45)$$

By substituting Eqs (8-43) and (8-45) into (8-44):

$$V_o = \frac{kV_xV_yR5}{R1} \qquad (8\text{-}46)$$

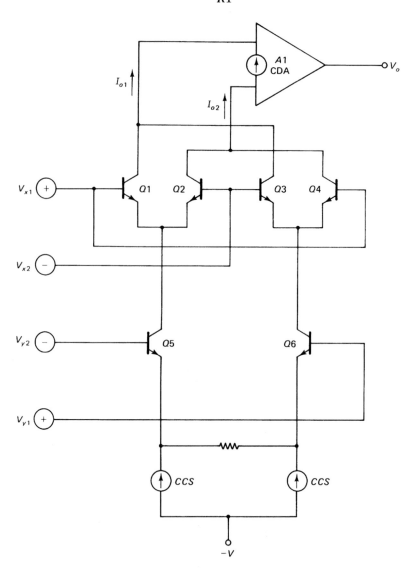

FIGURE 8-11 CDA multiplier circuit.

or, because $R5/R1$ is also a constant,

$$V_o = k_{tot} V_x V_y \qquad (8\text{-}47)$$

The transconductance cell forms the basis for most easily available IC multiplier–divider circuits. Figure 8-11 shows a typical IC multiplier circuit based on the transconductance cell. The output is a current, so it must be transformed into an output voltage in a *current difference amplifier* (CDA).

8-11 THE AD-533: A PRACTICAL ANALOG MULTIPLIER/DIVIDER

The Analog Devices, Inc. type AD-533 is a low-cost analog multiplier–divider circuit in integrated circuit form. Although these chips tend to be more expensive than the discrete circuits discussed above, they are also often more cost effective because they tend to work better and require less 'tweaking' to make them operate properly.

The pinouts for the 14-pin DIP version of the AD-533 is shown in Fig. 8-12. The AD-533 contains a transconductance multiplier for the X and Y inputs, and a summing junction at an operational amplifier for a Z input. The AD-533L version of this chip is capable of a full-scale linearity error of

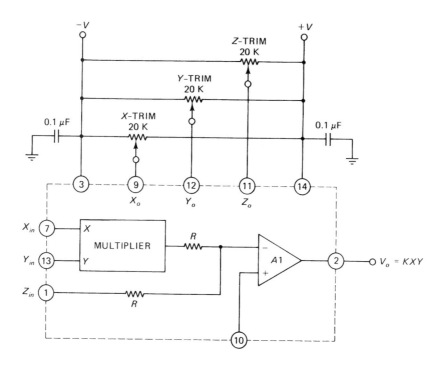

FIGURE 8-12 AD-533 multiplier.

only 0.5%. In addition, a low temperature coefficient of 0.01%/°C is provided. The AD-533 device is capable of operating to a small-signal bandwidth of 1 MHz, a full-power bandwidth of 750 kHz and a slew rate of 45 V/μs. The AD-533 will multiply in four quadrants with a transfer function of:

$$V_o = \frac{V_x V_y}{10} \text{ volts} \tag{8-48}$$

For division, the AD-533 will operate in two quadrants with a transfer function of:

$$V_o = \frac{10 V_z}{V_x} \text{ volts} \tag{8-49}$$

8-12 MULTIPLIER OPERATION OF THE AD-533

Figure 8-13 shows the connection of the AD-533 for straight multiplier operation. The V_x and V_y inputs are used for the operands, while the V_z input is fed back to the output. A scaling control circuit ($R1$ and $R2$) is used to set the output scale factor. The trimming circuits shown in Fig. 8-12 is also used, but is not shown here for the sake of simplicity. The range of V_x, V_y and V_o is ± 10 volts.

The trim procedure for the AD-533 multiplier is simple, and involves only three external multi-turn potentiometers. The procedure is also follows:

1. Set $V_x = V_y = 0$ volts, and then adjust Z-TRIM for $V_o = 0$ volts.
2. Set $V_x = 0$ volts, and then apply a signal of 50 Hz at 20 volts peak-to-peak; adjust X-TRIM for minimum signal output.
3. Set $V_y = 0$, and then apply a signal of 50 Hz at 20 volts peak-to-peak; adjust Y-TRIM for minimum signal output.

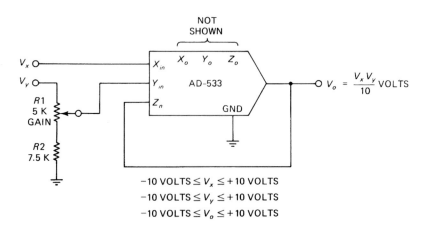

FIGURE 8-13 AD-533 multiplier circuit.

4. Readjust Z-TRIM for $V_o = 0$ Vdc output.
5. Set $V_x = +10$ volts, and V_y to 50 Hz at 20 volts peak-to-peak; adjust $R1$ for $V_o = V_y$.

Rearranging the connection of the AD-533 will render it a divider, squarer, squarerooter, and other function circuits.

8-13 DIVIDER OPERATION OF THE AD-533

The AD-533 can also be used as an analog divider circuit with the transfer function:

$$V_o = \frac{10V_z}{V_x} \text{ volts} \tag{8-50}$$

The circuit for the analog divider configuration is shown in Fig. 8-14. The X and Z inputs are used, with feedback going to the Y input. As was true in the previous case, the trimming circuits are not shown for sake of simplicity, but are used nonetheless. The trim procedure for the analog divider is as follows:

1. Set all trimmer potentiometers at mid-scale.
2. Set $V_z = 0$, adjust Z-TRIM so that V_o remains constant as V_x is varied from -10 Vdc through -1 Vdc.
3. With $V_z = 0$, $V_x = -10$ Vdc, set Y-TRIM for $V_o = 0$ Vdc.
4. Set $V_z = V_x$, and then adjust X-TRIM for the minimum variation of V_o as V_x is adjusted from -10 Vdc to -1 Vdc.
5. Repeat steps 2 and 3 for minimum interaction.

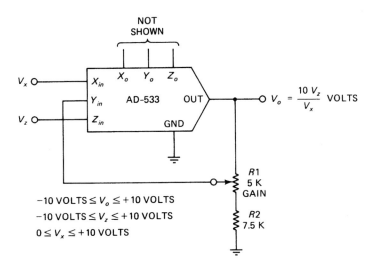

FIGURE 8-14 AD-533 divider circuit.

6. Set $V_z = V_x$, and then trim the gain control (R1) for V_o to have the closest approach to ±10 Vdc output as V_x is varied over the range −10 Vdc to −3 Vdc.

8-14 OPERATION OF THE AD-533 AS A SQUARER

A squarer is a circuit that outputs a voltage that is the square of the input voltage. Prior to the invention of good analog multiplier circuits the standard means for squaring a voltage in analog circuits was in a diode breakpoint generator. In those circuits, a series of diodes operated in the square law region of their characteristic are used to fit an output voltage to the square of its input voltage. While the method worked, it suffered from a severe temperature sensitivity, and also from the fact that only a limited number of breakpoints were possible. In the analog multiplier version of a squarer circuit, however, it is possible to simply apply the same voltage to both V_x and V_y inputs: if $V_x = V_y = V$, then $V_x V_y = V^2$.

Figure 8-15 shows a circuit for a squarer based on the AD-533 analog multiplier–divider. In this circuit, the Y_o input, which is normally connected to the Y-TRIM potentiometer, is grounded. The X-TRIM and Z-TRIM potentiometers are as shown earlier, but are deleted from Fig. 8-15 for simplicity sake. When properly trimmed, the output of the circuit in Fig. 8-15 is:

$$V_o = \frac{V_x^2}{10} \tag{8-51}$$

EXAMPLE 8-1
Find the output voltage of the squarer in Fig. 8-15 when the input voltage is −5 volts DC.

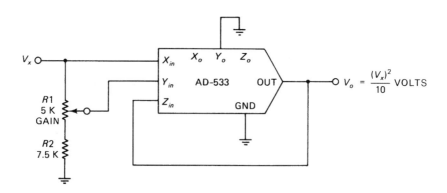

FIGURE 8-15 AD-533 squarer circuit.

Solution

$$V_o = \frac{V_x^2}{10}$$

$$V_o = \frac{(-5 \text{ volts})^2}{10}$$

$$V_o = \frac{25 \text{ volts}}{10} = 2.5 \text{ volts} \quad \blacksquare$$

The analog squarer circuit will operate in the two quadrants where the input voltage is positive and negative. The output voltage is, of course, only found in one quadrant (QI).

8-15 OPERATION OF THE AD-533 AS A FREQUENCY DOUBLER

There is a special case of the operation of the squarer that allows frequency doubling. When a sinewave signal is applied to the input of a squarer, the output frequency will be doubled. The AD-533 makes a nearly ideal frequency doubler because it does not require a tuned input or output network. The tuning circuits used on other analog frequency multipliers makes them single-frequency only. The frequency doubler based on the AD-533 is able to operate on any frequency within its range.

Consider the trigonometric identity:

$$\sin(A)\sin(B) = \frac{1}{2}[\cos(A - B) - \cos(A + B)] \quad (8\text{-}52)$$

If A and B each represent the same signal, then we can set $B = A$, and, then the identity takes the form:

$$\sin(A)\sin(A) = \frac{1}{2}[\cos(A - A) - \cos(A + A)] \quad (8\text{-}53)$$

or, collecting terms, and redefining A as $2\pi ft$:

$$\sin(2\pi ft)^2 = \frac{1}{2} - \frac{\cos(4\pi ft)}{2} \quad (8\text{-}54)$$

Rewriting Eq. (8-54) in the form that accounts for the standard IC multiplier which has a transfer function of:

$$V_o = \frac{V_x^2}{10} \quad (8\text{-}55)$$

and because $V_x = V\, 2\pi ft$:

$$V_o = \frac{(V\, 2\pi ft)^2}{20} \quad (8\text{-}56)$$

Using the identity of Eq. (8-54):

$$V_o = \frac{V^2}{20}[1 - \cos(4\pi f t)] \qquad (8\text{-}57)$$

EXAMPLE 8-2

A sinewave signal of 5 kHz at 10 volts peak is applied to the input of a squarer circuit. Calculate the output signal parameters.

Solution

(Note: the applied signal is defined at $10\sin(20\pi 5000 t)$)

1. The output of the squarer will be:

$$V_o = \frac{V_x^2}{10}$$

$$V_o = \frac{10(\sin 2\pi 5000 t)^2}{10}$$

$$V_o = \frac{[10\sin(2\pi 5000 t)]^2}{10}$$

$$V_o = \frac{100}{10}\sin(2\pi 5000 t)^2$$

2. The last expression is of the form suitable for using the trigonometric identity in Eq. (8-53):

$$V_o = 10\left[\frac{1}{2} - \frac{\cos(4\pi 5000 t)}{2}\right]$$

$$V_o = 5.00 - 5\cos(2\pi 10\,000 t)$$

The final expression shows a DC term of 5.00 volts and an AC term of $5\cos(2\pi 10\,000 t)$, or twice the input frequency. ∎

8-16 OPERATION AS A PHASE DETECTOR/METER

A multiplier driving a low-pass filter (Fig. 8-16) will operate as a phase meter. The output of this circuit is a DC level that is proportional to the difference in-phase (ϕ) between two different signals of the same frequency. Given a multiplier with a transfer function $V_o = V_x V_y/10$, and a low-pass filter that passed only signals with frequencies below the input frequency, only the DC level remains in the output, and that DC level is proportional to phase difference. The input signals to the circuit of Fig. 8-16 are:

$$V_x = V1\sin(2\pi f t) \qquad (8\text{-}58)$$

and

$$V_y = V2\sin(2\pi f t + \phi) \qquad (8\text{-}59)$$

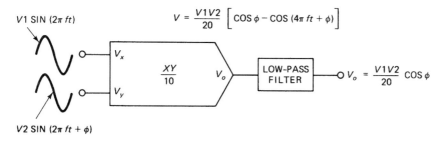

FIGURE 8-16 AD-533 phase detector.

Appealing to the identity of Eq. (8-52):

$$\sin(V_x)\sin(V_y) = \frac{1}{2}[\cos(V_x - V_y) - \cos(V_x + V_y)] \quad (8\text{-}60)$$

and for a multiplier with the transfer function $V_x V_y/10$, and substituting Eqs (8-58) and (8-59):

$$V_o = \frac{V1 V2}{20}[\cos(\phi) - \cos(4\pi f t + \phi)] \quad (8\text{-}61)$$

The expression of Eq. (8-61) represents the output of the analog multiplier. When the low-pass filter removes the AC term the equation becomes:

$$V_o = \frac{V1 V2}{20} \cos(\phi) \quad (8\text{-}62)$$

Solving Eq. (8-62) for the phase angle (ϕ) results in:

$$\phi = \cos^{-1}\left[\frac{20 V_o}{V1 V2}\right] \quad (8\text{-}63)$$

The phase detector operation described above assumes sinusoidal input signals. The circuit will also work with squarewaves or pulses, however, and in that case the output of the multiplier will be a series of pulses of width proportional to the phase difference. When the pulses are passed through the low-pass filter, the effect is to time average them and thereby produce a DC output that is zero when the signals are in quadrature, and maximum positive at 0° and maximum negative at 180° phase difference.

8-17 OPERATION OF THE AD-533 AS A SQUAREROOTER

A squarerooter is a circuit that will produce an output voltage that is the square root of the input voltage. Figure (8-17) shows the connection of the AD-533 as a squarerooter. In this circuit, the Y_o input is grounded, as was

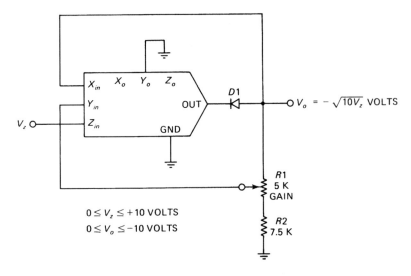

FIGURE 8-17 AD-533 squarerooter.

true with the squarer, but the X-TRIM and Z-TRIM external potentiometers are used (but not shown). The input signal is applied to the Z-input, while the Y and X inputs are fed back via a diode ($D1$). The circuit of Fig. 8-17 will accommodate input signals of 0 to +10 volts, but outputs a negative voltage in the range 0 to −10 volts. Therefore, the transfer function for this circuit is:

$$V_o = \sqrt{10V_z} \qquad (8\text{-}64)$$

If a positive output voltage is needed, then the output of Fig. 8-17 can be inverted in an operational amplifier circuit.

8-18 THE MULTIPLIER AS AN AUTOMATIC GAIN CONTROL (AGC)

An analog voltage multiplier can be used as an *automatic gain control* (AGC) if the right external circuitry is added. An AGC is a circuit that is designed to maintain a constant output signal level despite changes in the amplitude of the input signal. Examples of the use of AGC circuits include radio receivers and signal generators. The receiver uses the AGC (also sometimes called *automatic volume control* (or AVC) to maintain a level output signal despite the fact that various radio stations being received vary considerably in strength. A signal generator might use the AGC circuit to maintain a level output despite the fact that variable tuned oscillators tend to output different signal levels at different frequencies.

An AGC circuit based on an analog multiplier is shown in Fig. 8-18. In this circuit the multiplier acts essentially as a gain controlled amplifier. The

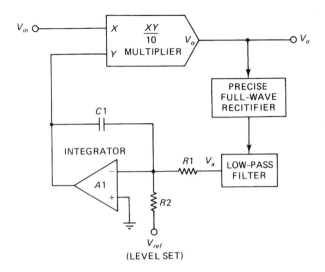

FIGURE 8-18 AD-533 automatic gain control circuit.

output of the multiplier is $V_o = V_x V_y/10$. The input signal is applied to the X input, while a DC feedback signal is applied to the Y input. The feedback signal is formed by sampling the output signal, V_o, rectifying it in a precise fullwave rectifier, and then filtering the rectifier output to form a DC signal (V_a) that is proportional to the output signal amplitude. This DC control signal is applied to the input of a Miller integrator where it is compared to a reference level (V_{ref}) that determines the set-point of the circuit. The integrator output is applied to the Y input of the multiplier. The Y input of the multiplier acts as a gain control signal. If an external level-set is needed, then the reference voltage V_{ref} can be made variable.

8-19 VOLTAGE CONTROLLED EXPONENTIATOR CIRCUIT

In Chapter 2, as part of the discussion of logarithmic amplifiers, the Burr-Brown 4301/4302 multifunction modules were discussed. These hybrid circuits have a transfer function of the form:

$$V_o = \left[\frac{V_y V_z}{V_x}\right]^m \quad (8\text{-}65)$$

In the transfer equation of the multi-function module (above) the exponent m is normally set by an external resistor voltage divider connected to the V_a, V_b and V_c inputs (see Fig. 8-19A). If $m = 1$, then the circuit operates as an ordinary multiplier or divider, depending upon which inputs are used. For example, when V_x is set to a constant reference of 10 volts, the multifunction module ($m = 1$) will operate as an $XY/10$ multiplier. Similarly, when V_z is

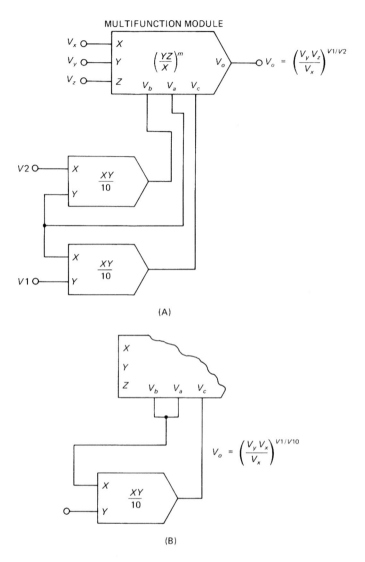

FIGURE 8-19 AD-533 exponentiator circuits.

set to a constant 10 volts, then the device functions as a divider of the form $10Y/X$. If $m < 1$, then the module operates as a rooter (e.g. a squarerooter when $m = 1/2$). If $m > 1$, then the module operates as an exponentiator, including a squarer. If m is set by a dynamic external voltage source, however, the multifunction module operates as a *voltage controlled exponentiator*.

In Fig. 8-19A the V_b and V_c inputs are driven by the outputs of a pair of $XY/10$ multipliers, which are in turn driven by control voltage $V1$ and $V2$. The V_a input is connected to the alternate inputs of the multipliers. The transfer function of the circuit in Fig. 8-19A is:

$$V_o = \left[\frac{V_y V_z}{V_x}\right]^{V1/V2} \tag{8-66}$$

A single control voltage version of the exponentiator circuit is shown in Fig. 8-19B (it is a partial version of Fig. 8-19A). V_a and V_b are connected together and applied to the X input of an external multiplier. The Y input of the multiplier becomes the control voltage, forming a transfer function of:

$$V_o = \left[\frac{V_y V_z}{V_x}\right]^{V1/10} \tag{8-67}$$

Similarly, if the output of the multiplier is connected to the V_b input, and V_a and V_c are connected to the X input of the multiplier, then the transfer function becomes:

$$V_o = \left[\frac{V_y V_z}{V_x}\right]^{10/V1} \tag{8-68}$$

8-20 SUMMARY

1. Multipliers have a transfer function KV_xV_y, while dividers have a transfer function KV_x/V_y. In most cases, $K = 1/10$.
2. Different forms of multipliers may operate in one, two or four quadrants.
3. Several different methods have been used for making multiplier circuits: matched resistor/lamp, matched JFET resistors, quarter square circuit, logarithmic amplifier, and transconductance cell. Of these, the log-amp and transconductance cells are the most commonly used today.
4. Typical applications of IC multiplier divider circuits include (in addition to multiplication and division): squarer, squarerooter, exponentiator, frequency doubler, phase detector/meter, and automatic gain control.

8-21 RECAPITULATION

Now return to the objectives and Pre-quiz questions at the beginning of the chapter and see how well you can answer them. If you cannot answer certain questions, place a check mark to each and review the appropriate parts of the text. Next, try to answer the questions and work the problems below, using the same procedure.

8-22 STUDENT EXERCISES

1. Design and build a transconductance analog multiplier based on an operational amplifier and a dual NPN transistor.
2. Design and build a quarter squares analog multiplier.

3. Connect and test the following AD-533 circuits: (a) multiplier, (b) divider, (c) squarer.

8-23 QUESTIONS AND PROBLEMS

1. List four different forms of analog multiplier circuit.
2. An analog multiplier has a scaling factor of 1/10. Calculate the output voltage when $V_x = +2.5$ volts and $V_y = +3$ volts.
3. An analog divider with a scaling factor of 1/10 sees the following input voltages: $V_x = +1.75$ volts, and $V_y = +5.00$ volts. Calculate the output voltage.
4. Draw the circuit for using an AD-533 as a phase detector/meter.
5. Demonstrate mathematically that the phase detector works.
6. Demonstrate mathematically that the squarer works as a frequency doubler when a sinewave signal is applied.
7. Find the output voltage of the squarer in Fig. 8-15 when the input voltage is -7.6 volts DC.
8. A sinewave signal of 10 kHz at 5 volts peak is applied to the input of a squarer circuit. Calculate the output signal parameters.
9. Write the basic equation for a quarter square multiplier, and evaluate it to prove that it can form a multiplier.
10. Draw the circuit for a logarithmic amplifier based multiplier, and define the voltages at the outputs of each stage.
11. Draw the circuit for a logarithmic amplifier based divider, and define the voltages at the outputs of each stage.
12. Sketch the circuit for an AD-533 connected as a multiplier circuit; as a divider.
13. Write the output expression for an analog multiplier when both inputs are connected together, and a sinewave of $10 \sin(2\pi f t)$ is applied.
14. An AD-533 is connected as a squarerooter. A +4 volt signal is applied to the input. What is the output voltage?

CHAPTER 9

Active filter circuits

OBJECTIVES

1. Recognize the response curves for low-pass, high-pass, bandpass, and notch filters.
2. Know the differences between Butterworth, Chebyshev, Cauer and Bessel filters.
3. Be able to design first, second and third-order filters.

9-1 PRE-QUIZ

These questions test your prior knowledge of the material in this chapter. Try answering them before you read the chapter. Look for the answers (especially those you answered incorrectly) as you read the text. After you have finished studying the chapter try answering these questions again, and those at the end of the chapter (see Section 9-13).

1. A _____ -pass filter passes only those frequencies above the cutoff frequency.
2. A cutoff slope of −6 dB/octave represents _____ dB/decade.
3. A _____ filter has a flat response within the passband and a heavily damped roll-off characteristic.
4. The _____ filter has a very rapid roll-off near the cutoff frequency, but only at the expense of less attenuation at frequencies far removed from the cutoff frequency.

9-2 INTRODUCTION TO ACTIVE FILTERS

An electronic frequency selective filter is a circuit that favors some frequencies and discriminates against other frequencies (or bands of frequencies). In other words, *a filter circuit will pass some frequencies and reject or sharply attenuate others*. Frequencies that pass through the filter with little attenuation are said to be in the *passband*, while frequencies that are heavily attenuated are said to be in the *stopband*. Filter circuits can be classified several ways: passive versus active, analog versus digital versus software, by frequency range (e.g. audio, RF or microwave), or by passband characteristic. *Passive filters* are made of various combinations of passive components such as resistors (R), capacitors (C) and inductors (L). In general, passive filters are lossy and not very flexible. An *active filter*, on the other hand, is based on an active device such as a transistor or an operational amplifier along with passive components (R, C and occasionally L) that determine frequency. In most cases, the passive components are resistors and capacitors (although a few inductor-based circuits are known).

Active filters use linear circuit techniques such as those found throughout this textbook. Digital filters use digital IC devices, and are often based on capacitor switching techniques. Software filters implement solutions to frequency selective equations using computer programming techniques. The emphasis in this chapter will be on analog active filters.

Filters can also be classified by frequency range. 'Audio' filters operate from the sub-audio to the ultrasonic range (near-DC to about 20 kHz). RF filters operate at frequencies above 20 kHz, up to about 900 MHz. Microwave filters operate at frequencies >900 MHz. These range designations are not absolute, but do serve to indicate approximate points at which a change of design techniques generally takes place. For example, filters can be made frequency selective using inductors. But in the audio range the inductance values are large, so inductors are bulky, costly and lossy. In addition, inductors produce stray magnetic fields that can interfere with other nearby circuits. On the other hand, inductors are the elements of choice in the RF region. But once frequencies approach several hundred megahertz, the inductance values required become too low for practical use so other techniques are required. In the microwave and high-UHF region transmission line and cavity techniques are used. The circuits discussed in this chapter are 'audio' active filters, but can have a passband between sub-audio and the low ultrasonic region.

Finally, filters may be classified by the nature of their frequency response characteristics. This method of categorizing filters takes note of the filter's passband and stopband. In this chapter we will examine *low-pass filters*, *high-pass filters*, *bandpass filters*, and *stopband filters*. We will also examine a related circuit called the *all-pass phase shifter*.

9-2.1 Filter characteristics

Figure 9-1A shows in general terms the characteristics of theoretically ideal filters. These curves will be discussed in greater detail later. A low-pass filter has a passband from DC to a specified cutoff frequency ($F1$). All frequencies above the cutoff frequency are attenuated, so are in the stopband. A bandpass filter has a passband between a lower limit ($F2$) and an upper limit ($F3$). All frequencies lower than $F2$, or greater than $F3$, are in the stopband. A high-pass filter has a stopband from DC to a certain lower limit ($F4$). All frequencies $>F4$ are in the passband.

A stopband filter response is shown in Fig. 9-1B. This filter severely attenuates frequencies between lower and upper limits ($F5$ to $F6$), but passes all others. When the stopband is very narrow, the stopband filter is called a *notch filter*. Such filters are often used to remove a single, unwanted frequency. An example of such an application is removal of unwanted

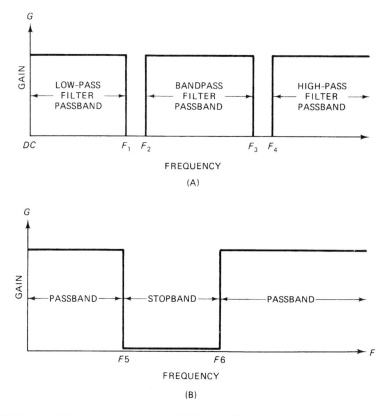

FIGURE 9-1 (A) Low-pass, bandpass and high-pass filter responses; (B) bandstop or notch filter response.

ACTIVE FILTER CIRCUITS

50/60 Hz interference (caused by proximity to local power lines) to medical electronic machines.

Ideal versus practical filter response curves. The response curves shown in Fig. 9-1 are unrealistic ideal generalizations. It might seem that a step-function cutoff is desirable, but in actual practice such a response is neither attainable nor desirable. The reason why it is undesirable is that the ideal response actually causes a problem because the filter may ring when a fast risetime signal is applied. This phenomena is especially likely to occur on narrow bandpass filters. While sinusoidal signals do not usually pose a problem, noise impulses or transient step functions can easily cause unwanted ringing.

There are several common filter responses: *Butterworth*, *Chebyshev*, *Cauer* (also sometimes called *elliptic*), and *Bessel*. The basic Butterworth response is shown in Fig. 9-2A. The noteworthy properties of the Butterworth filter is that both the passband and stopband are relatively flat, as is the transition region slope between them.

It is standard practice in filter design to specify the passband between points where the response falls off -3 dB from the mid-passband gain. Therefore, for the low-pass filter shown in Fig. 9-2A the cutoff frequency is at the point where gain ($A1$) falls off to 0.707 times the low frequency gain ($A1$), i.e. the -3 dB point.

At frequencies $f > f_c$, the gain falls off linearly at a rate that depends on the order of the filter. The slope (S) of the fall off is measured in either *decibels per octave* (a 2:1 frequency change) or *decibels per decade* (a 10:1 frequency change). Note that these two specifications can be scaled relative to each other: the -6 dB/octave roll-off has the same slope as -20 dB/decade. The slopes shown in Fig. 9-2A cover three Butterworth cases. A *first-order filter* offers a roll-off of -20 dB/decade, a *second-order filter* offers a roll-off of -40 dB/decade, and a *third-order filter* rolls off at -60 dB/decade. These correspond to 6, 12 and 18 dB/octave, respectively.

On first glance, it might appear that only third-order filters would be used because they transition from passband to stopband more rapidly. But higher order response is obtained at the cost of more complexity, greater sensitivity to component value error, and more difficult design. Some higher order filter designs are also more likely to oscillate than lower order equivalents. The selection of filter order is a trade-off between system requirements and complexity. If a very large, undesired signal is expected at a frequency up to, say, the third harmonic of f_c, then a higher order response might be indicated. Conversely, weaker stopband signals in that region may permit the use of a lesser order filter.

The steepness and shape of the roll-off curve is a function of the filter *damping factor*. As a class, Butterworth filters tend to be heavily damped, which explains the gradual roll-off in the response curve. The Chebyshev filter response (Fig. 9-2B) is lightly damped, so has a variation (or 'ripple')

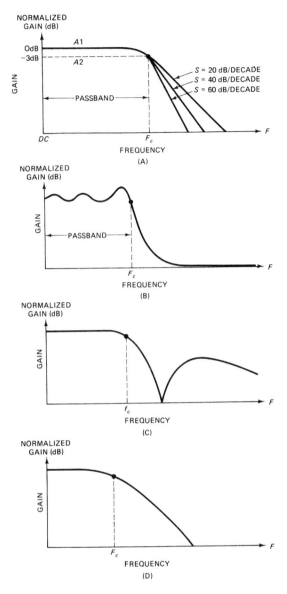

FIGURE 9-2 (A) Butterworth response; (B) Chebyshev response; (C) Cauer or elliptic response; (D) Bessel response.

within the passband. The Chebyshev filter offers a generally faster roll-off than the Butterworth filter, but at the cost of less flatness within the passband.

The Cauer or Elliptic filter response curve (Fig. 9-2C) offers the fastest roll-off for frequencies close to the cutoff frequency, as well as relatively

ACTIVE FILTER CIRCUITS **285**

while diode $D2$ is forward biased on positive peaks. Because $D1$ and $D2$ are shunted across $R3$, the total resistance $R3'$ is less than $R3$. By inspection of Eq. (10-82) one can determine that reducing $R3$ to $R3'$ reduces the gain of the circuit. The circuit is thus self-limiting.

Another variant of the gain-stabilized Wien bridge oscillator is shown in Fig. 10-28B. In this circuit a pair of back-to-back zener diodes provide the gain limitation function. With the resistor ratios shown the overall gain is limited to slightly more than unity, so the circuit will oscillate. The output peak voltage of this circuit is set by the zener voltages of $D1$ and $D2$ (which should be equal for low-distortion operation).

One final version of the gain stabilized oscillator is shown in Fig. 10-28C. In this circuit a small incandescent lamp is connected in series with resistor $R2$. When the amplitude of the output signal tries to increase above a certain level, the lamp will draw more current causing the gain to reduce. The lamp-stabilized circuit is probably the most popular form where stable outputs are required. A thermistor is sometimes substituted for the lamp.

10-7.3 Quadrature and biphasic oscillators

Signals that are *in quadrature* are of the same frequency but are phase shifted 90° with respect to each other. An example of quadrature signals is sine and cosine waves (Fig. 10-29A). Applications for the quadrature oscillator include demodulation of phase sensitive detector signals in data acquisition systems. The sinewave has an instantaneous voltage $v = V \sin(\omega_0 t)$, while the cosine wave is defined by $v = V \cos(\omega_0 t)$. Note that the distinction between sine and cosine waves is meaningless unless either both are present, or some other timing method is used to establish when 'zero degrees' is supposed to occur. Thus, when sine and cosine waves are called for it is in the context of both being present, and a phase shift of 90° is between them.

The circuit for the quadrature oscillator is shown in Fig. 10-29B. It consists of two operational amplifiers, $A1$ and $A2$. Both amplifiers are connected as Miller integrators, although $A1$ is a noninverting type while $A2$ is an inverting integrator. The output of $A1$ (i.e., V_{o1}) is assumed to be the sinewave output. In order to make this circuit operate, a total of 360° of phase shift is required between the output of $A1$, around the loop and back to the input of $A1$. Of the required 360° phase shift 180° are provided by the inversion inherent in the design of $A2$ (it is in the inverting configuration). Another 90° obtains from the fact that $A2$ is an integrator, which inherently causes a 90° phase shift. An additional 90° phase shift is provided by RC network $R3C3$. If $R1 = R2 = R3 = R$, and $C1 = C2 = C3 = C$, then the frequency of oscillation is given by:

$$f_{\text{Hz}} = \frac{1}{2\pi RC} \tag{10-85}$$

The Bessel filter also shows a phase shift over the passband, but it is nearly linear. A useful feature of this characteristic is that it allows a uniform time delay all across the passband. As a result, the Bessel filter offers the ability to pass transient pulse waveforms with minimum distortion. For the Bessel filter, the phase shift maximum is:

$$\Delta\phi_{max} = \frac{-n\pi}{2} \qquad (9\text{-}1)$$

where:
 $\Delta\phi_{max}$ is the maximum phase shift
 n is the order of the filter (i.e. number of poles)

In a properly designed Bessel filter, the cutoff frequency, f_c, occurs at a point where the phase shift is half the maximum phase shift, or:

$$\Delta\phi_{fc} = \frac{-n\pi}{4} \qquad (9\text{-}2)$$

The Bessel filter is said to work best at the frequency where $f = f_c/2$.

9-3 LOW-PASS FILTERS

In this section, and the sections to follow, practical design of active filters will be discussed. The model for the filter is shown in Fig. 9-4. This filter is called the *voltage controlled voltage source* (VCVS) filter or *Sallen-Key* filter. The basic configuration is a noninverting follower operational amplifier (A1). The op-amp selected should have a high gain–bandwidth product, relative to the cutoff frequency, in order to permit the filter to operate properly. The gain

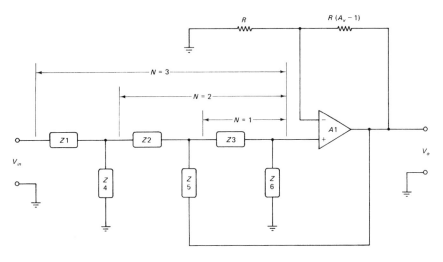

FIGURE 9-4 Voltage controlled voltage source (VCVS), or Sallen-Key filter generic circuit.

of the circuit is given by:

$$A_v = \frac{R_f}{R_{in}} + 1 \qquad (9\text{-}3)$$

so, if $R_{in} = R$, then we may deduce:

$$R_f = R(A_v - 1) \qquad (9\text{-}4)$$

In some circuits, the gain may be unity. In those circuits the resistor voltage divider feedback network is replaced with a single connection between output and the inverting input.

The input circuitry of the generic VCVS filter consists of a network of impedances labelled Z1 through Z6. Each of these blocks will be either a resistance (R) or a complex capacitive reactance ($-jX_c$). Which element becomes which type of component is determined by whether the filter is a low-pass or high-pass type.

The order of the filter, denoted by n, refers to the number of poles in the design, or in practical terms, the number of RC sections. A first-order filter ($n = 1$) consists of Z3 and Z6, a second-order filter ($n = 2$) consists of Z2, Z3, Z5 and Z6, and a third-order filter ($n = 3$) consists of all six impedances (Z1–Z6). Higher order filters ($n > 3$) can also be built, but are not discussed here.

In a low-pass filter Z1 through Z3 are resistances, while Z4 through Z6 are capacitances. The component roles are reversed in high-pass filters. Now that we have laid a foundation, let's review the properties of the low-pass filter and then learn to design first, second and third-order low-pass VCVS filters.

By way of review, Fig. 9-5 shows the low-pass Butterworth filter response curve. This type of filter is maximally flat within the passband, and passes all

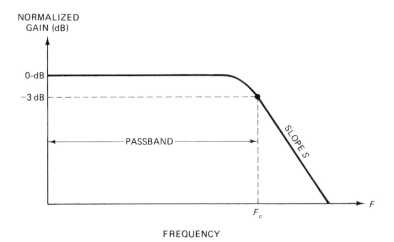

FIGURE 9-5 Frequency response curve.

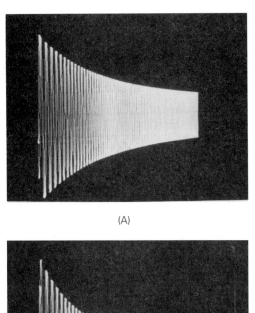

(A)

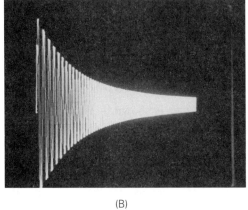

(B)

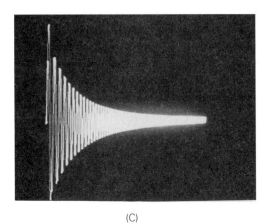

(C)

FIGURE 9-6 Frequency sweeps of low-pass filters: (A) first-order; (B) second-order; (C) third-order.

frequencies below a certain critical frequency (F_c). The breakpoint between the passband and the stopband is the point at which the gain of the circuit has dropped off -3 dB from its lower frequency value. Above the critical frequency the gain falls off at a certain rate indicated by the slope of the curve. The steepness of the slope is usually specified in terms of decibels (dB) of gain per octave of frequency or, alternatively, dB/decade is sometimes used.

Frequency sweeps of a set of low-pass filters are shown in Fig. 9-6. These oscilloscope traces were created using a sweep generator that varies a sinewave signal from 1 kHz to 10 kHz. The -3 dB frequency of the filters was 3 kHz. The Y-input of the oscilloscope displays the output amplitude of the filter, while the X-input of the 'scope was externally swept using the same sawtooth that was used to sweep the signal generator. Thus, the X–Y oscilloscope trace is a plot of the frequency response of the filter under test. Figure 9-6A shows the response of a first-order filter in the area from just below f_c up to 10 kHz. The response of the second-order filter is shown in Fig. 9-6B, and for the third-order filter in Fig. 9-6C. All three traces were taken under the same conditions, so one can see how the attenuation of frequencies greater than f_c is greater in the higher order filters.

9-3.1 First-order low-pass (−20 dB/decade) filters

The first-order low-pass filter is shown in Fig. 9-7A, and its response curve is shown in Fig. 9-7B. The filter consists of a single-section RC low-pass filter driving the noninverting input of an operational amplifier. The gain of the op-amp is $[(R2/R3) + 1]$. The high input impedance of $A1$ prevents loading of the RC network. The general form of the transfer equation for the amplitude versus frequency response for the first-order filter is:

$$A_{dv} = 20 \log A_v - 20 \log \sqrt{1 + \omega_o^2} \qquad (9\text{-}5)$$

where:

    A_{dB} is the gain of the circuit in decibels
    A_v is the voltage gain within the passband
    log denotes the base-10 logarithms
    ω_o is the ratio of the input frequency to the cutoff frequency
    ($f_o = f/f_c$)

The voltage at the output of the RC network (V_a) is found from the voltage divider equation:

$$V_a = \frac{-jX_c V_{in}}{R - jX_c} \qquad (9\text{-}6)$$

where: $-jX_c = 1/j2\pi fC$ and j is the imaginary operator

$$j = \sqrt{-1}$$

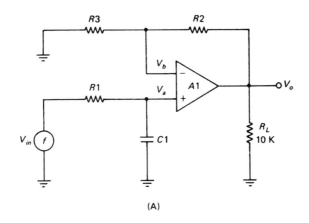

(A)

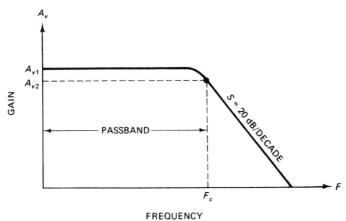

$$A_{v1} = \frac{R2}{R3} + 1$$

$$A_{v2} = A_{v1} - 3\,\text{dB} = 0.707\,A_{v1}$$

(B)

FIGURE 9-7 (A) First-order low-pass filter; (B) response.

Substituting the value for $-jX_c$:

$$V_a = \frac{\dfrac{V_{in}}{j2\pi fC}}{R + \dfrac{1}{j2\pi fC}} \qquad (9\text{-}7)$$

which simplifies to:

$$V_a = \frac{V_{in}}{1 + 2\pi FCR} \qquad (9\text{-}8)$$

If the transfer function of the noninverting follower is:

$$V_o = V_{in}\left[\frac{R2}{R3} + 1\right] \quad (9\text{-}9)$$

and since $V_{in} = V_a$ (Eq. (9-7)):

$$V_o = \left[\frac{V_{in}}{1 + 2\pi f CR}\right] \times \left[\frac{R2}{R3} + 1\right] \quad (9\text{-}10)$$

Equation (9-9) can be put into a more general transfer equation of the form:

$$\frac{V_o}{V_{in}} = \frac{A_v}{1 + j(f/f_c)} \quad (9\text{-}11)$$

where:

V_o is the output signal voltage
V_{in} is the input signal voltage
A_v is the passband gain $[(R2/R3) + 1]$ (see Eq. (9-8))
f is the signal frequency in hertz (Hz)
f_c is the -3 dB frequency $(1/2\pi RC)$

The filter parameters are required to define the operation of any particular circuit. The *gain magnitude* and *phase shift* are found from the following equations:

Gain magnitude

$$\left|\frac{V_o}{V_{in}}\right| = \frac{A_v}{\sqrt{1 + (f/f_c)^2}} \quad (9\text{-}12)$$

and phase shift angle (in radians):

$$\phi = -\tan^{-1}(f/f_c) \quad (9\text{-}13)$$

EXAMPLE 9-1

A first-order low-pass VCVS filter is built with the following component values: $R1 = 10$ kohms, $R2 = 2$ kohms and $R3 = 1$ kohm, $C1 = 0.01$ µF. Characterize this filter if a 1200 Hz sinewave signal is applied.

Solution

1. Calculate the -3 dB frequency:

$$f_c = \frac{-1}{2\pi R1 C1}$$

$$f_c = \frac{-1}{(2)(3.14)(10\,000 \text{ ohms})(0.01 \times 10^{-6} \text{ F})} = 1592 \text{ Hz}$$

2. Passband gain (A_v):

$$A_v = \frac{R2}{R3} + 1$$

$$A_v = \frac{2\ \text{k}\Omega}{1\ \text{k}\Omega} + 1 = 3$$

3. Calculate gain magnitude:

$$\left|\frac{V_o}{V_{in}}\right| = \frac{A_v}{\sqrt{1 + (f/f_c)^2}}$$

$$\left|\frac{V_o}{V_{in}}\right| = \frac{3}{\sqrt{1 + (1200/1592)^2}}$$

$$\left|\frac{V_o}{V_{in}}\right| = \frac{3}{\sqrt{1 + 0.057}}$$

$$\left|\frac{V_o}{V_{in}}\right| = \frac{3}{1.25} = 2.4$$

4. Phase shift angle (in radians):

$$\phi = -\tan^{-1}(f/f_c)$$

$$\phi = -\tan^{-1}(1200/1592)$$

$$\phi = -\tan^{-1}(0.75) = 36.87°$$ ■

Because the filter characterization depends in part on the ratio f/f_c, the equations take different forms at different values of f and f_c. These can be reduced to:

At low frequencies well within the passband ($f < f_c$):

$$\left|\frac{V_o}{V_{in}}\right| = A_v = \frac{R2}{R3} + 1 \qquad (9\text{-}14)$$

At the −3 dB cutoff frequency ($f = f_c$):

$$\left|\frac{V_o}{V_{in}}\right| = 0.707 A_v \qquad (9\text{-}15)$$

At a high frequency well above the −3 dB cutoff frequency ($f > f_c$):}

$$\left|\frac{V_o}{V_{in}}\right| < A_v \qquad (9\text{-}16)$$

TABLE 9-1 Filter characteristics.

Applied frequency (Hz)	Gain magnitude A_v	dB	Phase shift (degrees)
10	2.00	6.02	−0.57
20	1.999	6.018	−1.15
50	1.998	6.009	−2.86
80	1.994	5.993	−4.57
100	1.990	5.977	−5.71
200	1.961	5.850	−11.31
500	1.789	5.052	−26.57
800	1.561	3.872	−38.66
1000	1.414	3.010	−45
2000	0.894	−0.969	−63.43
5000	0.392	−8.129	−78.69
8000	0.248	−12.109	−82.87
10 000	0.199	−14.023	−84.29
20 000	0.0998	−20.011	−87.14
50 000	0.0399	−27.960	−88.85
80 000	0.0249	−32.042	−89.28
100 000	0.0199	−33.980	−89.43

Table 9-1 shows the characteristics of first-order filters at several different ratios of f/f_c.

9-3.2 Design procedure for a first-order low-pass filter

There are two basic ways to design a low-pass filter: *ground-up* and by *frequency scaling*. In this section the ground-up method is discussed.

Procedure
1. Select the −3 dB cutoff frequency (f_c) from consideration of the circuit requirements and applications.
2. Select a standard value capacitance ($C \leq 1$ µF) as a trial.
3. Calculate the required resistance from:

$$R1 = \frac{1}{2\pi f_c C}$$

4. Select the passband gain for $f < f_c$.
5. Select a value for resistor R, and
6. Calculate R_f from:

$$R_f = R(A_v - 1)$$

EXAMPLE 9-2
A low-pass filter is needed for a biological fluid pressure transducer. The cutoff frequency should be 100 Hz, and the gain should be 5.

Solution

1. $F_c = 100$ Hz
2. Select trial value for C1: 0.1 µF
3. Calculate R1:

$$R1 = \frac{1}{2\pi f_c C1}$$

$$R1 = \frac{1}{(2)(3.14)(100 \text{ Hz})(0.1 \times 10^{-6} \text{ µF})}$$

$$R1 = \frac{1}{6.28 \times 10^{-5}} \text{ ohms} = 15923 \text{ ohms}$$

4. Select a trial value for R1: 10 kohms
5. Calculate R_f:

$$R_f = R(A_v - 1)$$
$$R_f = (10 \text{ k}\Omega)(5 - 1)$$
$$R_f = (10 \text{ k}\Omega)(4) = 40 \text{ k}\Omega$$ ∎

9-3.3 Design by frequency normalized model

The filter design can be simplified by using a normalized model. First, design a filter for a standardized frequency (e.g. 1 Hz, 10 Hz, 100 Hz, 1000 Hz or 10 000 Hz) and list the component values. The values for any other frequency can then be computed for any other frequency by a simple ratio and proportion. An example of a first-order low-pass Butterworth filter is shown in Fig. 9-8. The component values shown in Fig. 9-8 are normalized for 1 kHz.

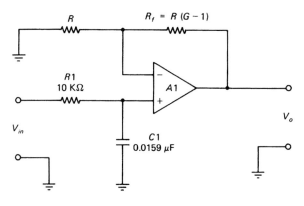

F_c NORMALIZED TO 1-KHz

FIGURE 9-8 Normalized first-order low-pass filter.

The actual required component values ($R1'$ and $C1'$) are found by dividing the normalized values shown by the desired cutoff frequency in kilohertz.

$$C1' = \frac{(C1)(1 \text{ kHz})}{F} \qquad (9\text{-}17)$$

or

$$R1' = \frac{(R1)(1 \text{ kHz})}{F} \qquad (9\text{-}18)$$

Leave one of the values alone, and calculate the other. In general, it is easier to obtain precision resistors in unusual values (or the value obtained by a potentiometer), so it is common practice to select a standard capacitance and calculate the new resistance.

EXAMPLE 9-3

Change the frequency of the normalized 1 kHz filter to 60 Hz (i.e. 0.06 kHz).

Solution

$$C1' = \frac{(C1)(1 \text{ kHz})}{F} \qquad \text{Eq. (9-17) restated}$$

$$C1' = \frac{(0.0159)(1 \text{ kHz})}{0.06} = 0.265 \text{ μF} \qquad \blacksquare$$

A generalized form of the equations is given below:

$$A' = A(f/f_c) \qquad (9\text{-}19)$$

where:
- A' is the new component value for either $C1$ or $R1$
- A is the original component value for either $C1$ or $R1$
- f_c is the filter -3 dB cutoff frequency
- f is the new design frequency

9-3.4 Second-order low-pass (−40 dB/decade) filters

The circuit for a second-order low-pass filter is shown in Fig. 9-9A, while the response curve is shown in Fig. 9-9B. Note that this circuit is similar to the first-order filter, but with an additional RC network in the frequency selective portion of the circuit.

The particular version of this circuit shown in Fig. 9-9A is connected in the unity gain configuration. The purpose of $R3$ is to help counteract the DC offset at the output of the operational amplifier that is created by input bias currents charging the capacitors in the frequency selective network. The value of $R3$ in the unity gain case is $2R$, where R is the value of the resistors in the frequency selective network. In cases where DC offset is not a problem, resistor $R3$ can be replaced with a short circuit between the op-amp output

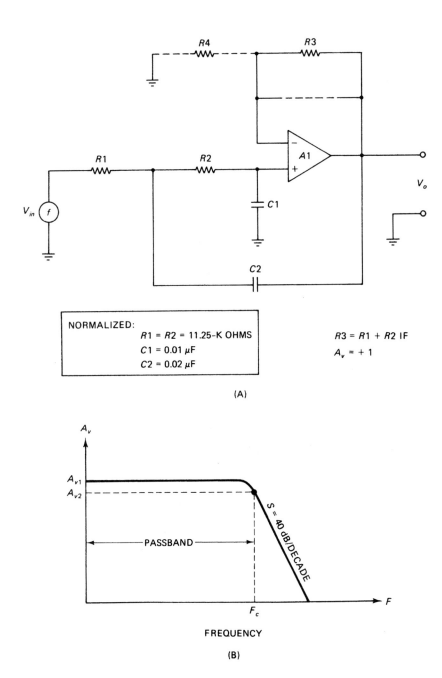

FIGURE 9-9 Second-order low-pass filter: (A) circuit; (B) response.

and the inverting input. If passband gain is required, then resistors $R3$ and $R4$ are used.

The second-order VCVS filter is by far the most commonly used type. Its -40 dB/decade roll-off, coupled with a high degree of stability, results in a generally good trade-off between performance and complexity.

The general form of the second-order filter transfer equation is similar to the expression for the first-order filter:

$$A_{dB} = 20 \log A_v - 20 \log \sqrt{\omega_o^4 + (a^2 - 2)(\omega_o^2) + 1} \qquad (9\text{-}20)$$

[Note: a is the *damping factor* of the circuit, and other terms are as defined earlier for the first-order case.]

The damping factor term (a) is determined by the form of filter circuit. For the Butterworth design, which is used in most of the examples of filters in this chapter, the value of a is $\sqrt{2}$, or 1.414.

The passband gain for this circuit is the normal gain for any noninverting follower/amplifier. If the output is strapped directly to the inverting input, or if $R3$ (but not $R4$) is used in the feedback network, then the gain is unity ($A_v = +1$). For gains greater than unity ($A_v > 1$), the following is true:

$$A_v = \frac{R3}{R2} + 1 \qquad (9\text{-}21)$$

The cutoff frequency (f_c) is the frequency at which the voltage gain drops -3 dB from the passband gain (Eq. (9-21)). This gain is found from:

$$A_v = \frac{1}{2\pi \sqrt{R1R2C1C2}} \qquad (9\text{-}22)$$

The gain magnitude (abs(V_o/V_{in})) is found in a manner similar to the first-order case:

$$\left| \frac{V_o}{V_{in}} \right| = \frac{A_v}{\sqrt{1 + (f/f_c^4)}} \qquad (9\text{-}23)$$

There is no requirement in VCVS filters that like components (R or C) in the frequency selective network be made equal, but such a step simplifies the design procedure. If $R1 = R2 = R$, and $C1 = C2 = C$, then:

$$f_c = \frac{1}{2\pi f_c C} \qquad (9\text{-}24)$$

A constraint on this simplification is that the Butterworth response is guaranteed only if $A_v \leq 1.586$.

9-3.5 Design procedure

1. Select the -3 dB cutoff frequency (f_c) from consideration of the circuit requirements and applications.

2. Select a standard value capacitance (30 pF ≤ C ≤ 1 µF) as a trial value.
3. Calculate the required resistance from:

$$R1 = \frac{1}{2\pi f_c C} \qquad (9\text{-}25)$$

4. Select the passband gain for $f < f_c$.
5. Select a value for resistor $R4$, and
6. Calculate $R3$ from:

$$R3 = R4(A_v - 1) \qquad (9\text{-}26)$$

EXAMPLE 9-4

Design a 1 kHz second-order low-pass filter with a unity gain based on Fig. 9-9.

Solution

1. Select f_c: 1 kHz (given)
2. Select a trial value for $C1$ and $C2$: $C = 0.0056$ µF
3. Calculate $R1 = R2 = R$:

$$R = \frac{1}{2\pi f_c C}$$

$$R = \frac{1}{(2)(3.14)(1000 \text{ Hz})(0.0056 \times 10^{-6} \text{ farads})}$$

$$R = \frac{1}{3.52 \times 10^{-5}} \text{ ohms} = 28\,435$$

4. Select $R3$:

$$R3 = 2R = (2)(28\,435 \text{ ohms}) = 56\,870 \text{ ohms} \qquad \blacksquare$$

The normalized 1 kHz trial values for doing scaling design of the second-order low-pass filter are shown in Fig. 9-9 as an inset. The design here is based on a more complex arrangement whereby $C2 = 2C1$. Some authorities maintain that this is the superior design. The same scaling rule is applied to the second-order filter as was used in the first-order.

EXAMPLE 9-5

Design a 3000 Hz second-order low-pass filter with a gain of one using the scaling technique.

Solution

$$R1' = \frac{R1 f_c}{f}$$

$$R1' = \frac{(11.25 \text{ k}\Omega)(1000 \text{ Hz})}{3000 \text{ Hz}} = 3.75 \text{ k}\Omega \qquad \blacksquare$$

9-3.6 Third-order low-pass (−60 dB/decade) filters

A third-order filter has a frequency roll-off slope of −60 dB/decade, or −18 dB/octave. There are two main forms of third-order filter. One type is similar to the first and second-order filters, except for an extra low-pass *RC* filter in the frequency selective network. The other method is to cascade first and second-order filters. Figure 9-10A shows the circuit for the former type, along with the response curve shown in Fig. 9-10B. This circuit is the normalized version. In past examples, filters were normalized using frequency

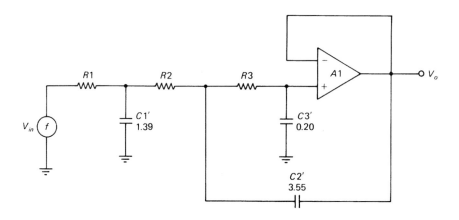

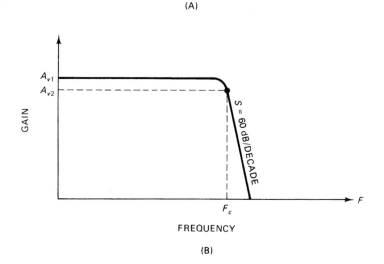

FIGURE 9-10 Normalized second-order low-pass filter and response.

300 ACTIVE FILTER CIRCUITS

f_c as the determining factor, so in this example we will use the radian form in which frequency ϖ_c is the cutoff frequency in radians per second, and is equal to $2\pi f_c$. The filter circuit shown in Fig. 9-10A is normalized to one radian per second.

The values assigned to the capacitor values in Fig. 9-10A are parametric values, and are used in a two-step process for finding the actual final values C1, C2 and C3. The parameters are $C1' = 1.39$, $C2' = 3.55$, and $C3' = 0.20$ (for Butterworth filters).

EXAMPLE 9-6

Design a -60 dB/octave low-pass filter for a frequency of 100 Hz. Use the scaling method, and the normalized values shown in Fig. 9-10A.

Solution

1. Calculate ϖ_c:

$$\varpi_c = (100 \text{ Hz})(2)(3.14) = 628 \text{ radians/second}$$

2. State the parameter values:

$$C1' = 1.39$$
$$C2' = 3.55$$
$$C3' = 0.20$$

3. Perform frequency scaling:

$$C1a = \frac{C1}{\varpi_c} = \frac{1.39}{628} = 2.2 \times 10^{-3}$$

$$C2a = \frac{C2}{\varpi_c} = \frac{3.55}{628} = 5.65 \times 10^{-3}$$

$$C3a = \frac{C31}{\varpi_c} = \frac{0.20}{628} = 3.19 \times 10^{-4}$$

4. Select a trial value for C2: 0.1 µF (e.g. 1×10^{-7} farads).
5. Solve for the value of $R1 = R2 = R3 = R$:

$$R = \frac{C2a}{C2} = \frac{5.65 \times 10^{-3}}{10^{-7}} = 56\,500 \text{ ohms}$$

$$C1 = \frac{C1a}{R} = \frac{2.2 \times 10^{-3}}{56\,500} = 3.9 \times 10^{-8} = 0.0039 \text{ µF}$$

$$C3 = \frac{C3a}{R} = \frac{3.19 \times 10^{-4}}{56\,500} = 5.7 \times 10^{-9} = 0.0057 \text{ µF}$$ ∎

A second method for designing a third-order low-pass filter is to cascade first-order and second-order filters of the same cutoff frequency (Fig. 9-11).

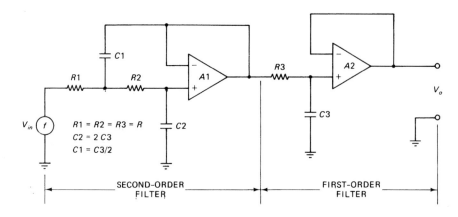

FIGURE 9-11 Third-order low-pass filter.

The roll-off of the first-order filter is -20 dB/decade, and for the second-order filter it is -40 dB/decade. The combined roll-off is $-(20+40) = -60$ dB/decade. The components in this circuit have the following relationships:

$$R1 = R2 = R3 = R = \frac{1\,000\,000}{2\pi f_c C3_{\mu F}} \qquad (9\text{-}27)$$

$C3$ is selected for convenient value:

$$C1 = \frac{C3}{2} \qquad (9\text{-}28)$$

$$C2 = 2C3 \qquad (9\text{-}29)$$

The design procedure calls for selecting a reasonable trial value for $C3$, and then calculating $C1$ and $C2$. If these values are not standards, then set another trial value of $C3$ and try again. Repeat the procedure until a good set of values if obtained. Finally, calculate R.

EXAMPLE 9-7

Design a -60 dB/decade cascade filter for a frequency of 500 Hz.

Solution
1. Select frequency f_c: 500 Hz
2. Select $C3$: 0.01 µF
3. Calculate $C1$:
$$C1 = C3/2 = 0.01\ \mu F/2 = 0.005\ \mu F$$
4. Calculate $C2$:
$$C2 = 2 \times C3 = (2)(0.01\ \mu F) = 0.02\ \mu F$$

5. Calculate $R1 = R2 = R3 = R$:

$$R = \frac{1\,000\,000}{2\pi f_c C3_{\mu F}}$$

$$R = \frac{1\,000\,000}{(2)(3.14)(500 \text{ Hz})(0.01 \text{ }\mu F)}$$

$$R = \frac{1\,000\,000}{31.4} = 31\,850 \text{ ohms} \qquad \blacksquare$$

9-4 HIGH-PASS FILTERS

The high-pass filter is the inverse of the low-pass filter, so one can reasonably expect its frequency response characteristic to mirror that of the low-pass filter response. Figure 9-12 shows the basic high-pass filter response with roll-off slopes of -20, -40 and -60 dB/decade. The passband of the high-pass filter are all frequencies above the cutoff frequency f_c. As in the low-pass case, f_c is the frequency at which passband gain drops -3 dB; that is $A_{vc} = 0.707 A_v$.

The cutoff frequency phase shift in a high-pass filter has the same magnitude as the low-pass case, but the sign is opposite. At f_c, the high-pass filter exhibits a phase shift of $+45°$ per 20 dB/decade of roll-off. Put another way, the phase shift is $(n \times 45°)$, where n is the order of the filter.

The high-pass versions of the VCVS filters are of the same form as the low-pass filter. That form was laid out in Fig. 9-4 earlier. In the case of the high-pass filter, however, impedances Z1 through Z3 are capacitances, while Z4 through Z6 are resistances. In the high-pass filter the roles of the resistors and capacitors are reversed.

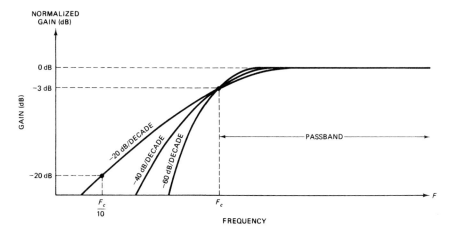

FIGURE 9-12 Responses for first, second and third-order high-pass filters.

9-4.1 First-order high-pass VCVS filters (−20 dB/decade)

The circuit for a first-order high-pass filter is shown in Fig. 9-13. This circuit is identical to the first-order low-pass filter in which the roles of $C1$ and $R1$ are interchanged. The filter shown here is normalized for 1 kHz. Passband gain of this circuit is:

$$A_v = \frac{R_f}{R_{in}} \quad (9\text{-}30)$$

The voltage at the noninverting input of the operational amplifier (V_a) is developed across resistor $R1$, and is given by:

$$V_a = \frac{j2\pi f R1 C1 V_{in}}{1 + j2\pi f R1 C1} \quad (9\text{-}31)$$

The transfer equation for the circuit is:

$$V_o = A_v V_a \quad (9\text{-}32)$$

$$V_o = \left[\frac{R_f}{R_{in}} + 1\right]\left[\frac{j2\pi f R1 C1 V_{in}}{1 + j2\pi f R1 C1}\right] \quad (9\text{-}33)$$

and, in the traditional form, the equation becomes:

$$\frac{V_o}{V_{in}} = \frac{A_v j(f/f_c)}{1 + j(f/f_c)} \quad (9\text{-}34)$$

As in the previous cases, the cutoff frequency f_c is found from:

$$f_c = \frac{1}{2\pi R1 C1} \quad (9\text{-}35)$$

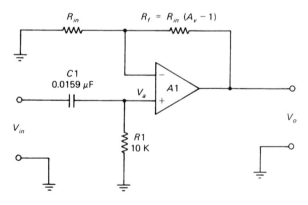

FIGURE 9-13 First-order high-pass filter.

The gain magnitude of this circuit is the absolute value of the traditional form of the transfer equation:

$$\left|\frac{V_o}{V_{in}}\right| = \frac{A_v(f/f_c)}{\sqrt{1+(f/f_c)^2}} \quad (9\text{-}36)$$

The VCVS high-pass filter shown in Fig. 9-12 is normalized to 1 kHz. The same scaling technique is used for this circuit as was used for the low-pass filters discussed earlier.

9-4.2 Second-order high-pass filter (−40 dB/decade)

The second-order high-pass filter (Fig. 9-14) offers a roll-off slope of −40 dB/decade. This VCVS filter circuit is, like its low-pass counterpart, probably the most commonly used form of high-pass filter. The circuit is similar to the low-pass design except for a reversal of the roles of capacitors and resistors. The cutoff frequency is the frequency at which gain falls off −3 dB, and is found from:

$$f_c = \frac{1}{2\pi\sqrt{R1R2C1C2}} \quad (9\text{-}37)$$

or, in the case where $R1 = R2 = R$, and $C1 = C2 = C$:

$$f_c = \frac{1}{2\pi RC} \quad (9\text{-}38)$$

The gain magnitude of the circuit is found from:

$$\left|\frac{V_o}{V_{in}}\right| = \frac{A_v}{\sqrt{1+(f/f_c)^4}} \quad (9\text{-}39)$$

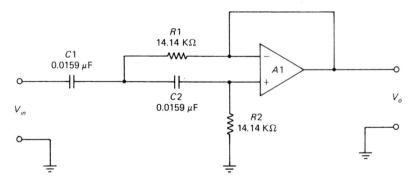

FIGURE 9-14 Second-order high-pass filter.

EXAMPLE 9-8

Calculate the cutoff frequency of a filter such as Fig. 9-14 in which $C1 = C2 = 0.0056\ \mu F$ and $R1 = R2 = 22$ kohms.

Solution

$$f_c = \frac{1}{2\pi RC}$$

$$f_c = \frac{1}{(2)(3.14)(22\,000\ \Omega)(0.0056 \times 10^{-6}\ \text{farads})}$$

$$f_c = \frac{1}{7.74 \times 10^{-4}} = 1.293\ \text{Hz} \qquad \blacksquare$$

9-4.3 Multiple feedback path filters

A different form of active filter, the *multiple feedback path* (MFP) circuit, is shown in Fig. 9-15. The low-pass version is shown in Fig. 9-15A and the

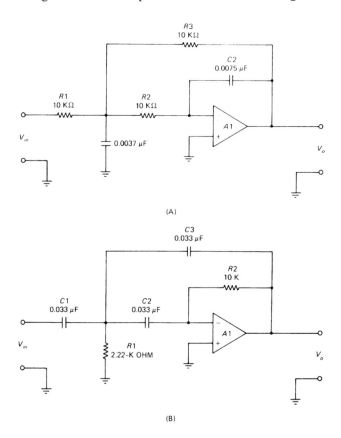

FIGURE 9-15 Multiple feedback path filters: (A) low-pass; (B) high-pass.

high-pass in Fig. 9-15B. The values are normalized for 1 kHz, and we find the actual values in the manner described above. Change either the capacitor or the resistor values, but not both.

9-5 BANDPASS FILTERS

The bandpass filter is a circuit that has a passband between an upper limit and a lower limit. Frequencies above and below these limits are in the stopband. There are two basic forms of bandpass filter: *wide bandpass* and *narrow bandpass*. These two types are sufficiently different that they offer different

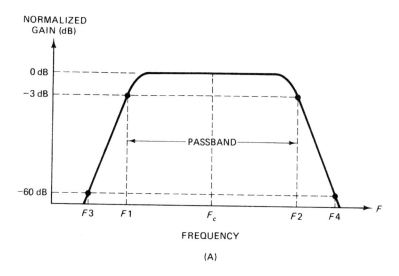

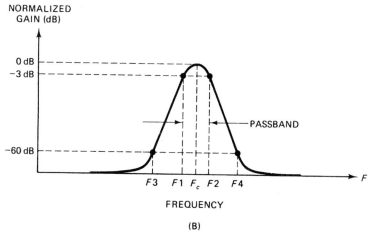

FIGURE 9-16 (A) Low-Q bandpass filter; (B) high-Q bandpass filter.

responses. The wide bandpass filter may have a passband that is wide enough to be called a bandpass amplifier rather than a filter. The wide bandpass filter response is shown in Fig. 9-16A, while the narrow response is shown in Fig. 9-16B. The passband is defined as the frequency difference between the upper −3 dB point ($F2$), and the lower −3 dB point ($F1$). The bandwidth BW is:

$$BW = F2 - F1 \qquad (9\text{-}40)$$

The center frequency f_c of the bandpass filter is usually symmetrically placed between $F1$ and $F2$, or $(F2 - F1)/2$. If the filter is a very wideband type, however, the center frequency is:

$$f_c = \sqrt{F1F2} \qquad (9\text{-}41)$$

Bandpass filters are sometimes characterized by the *figure of merit*, or Q. The Q is a factor that describes the sharpness of the filter, and is found from:

$$Q = \frac{f_c}{BW} \qquad (9\text{-}42)$$

$$Q = \frac{f_c}{F2 - F1} \qquad (9\text{-}43)$$

The Q of the filter tells us something of the passband characteristic. Wideband filters generally have a $Q < 10$, while narrow band filter have a $Q > 10$.

EXAMPLE 9-9

Determine the Q of a filter that has a center frequency of 2000 Hz, and −3 dB frequencies of 1800 Hz and 2200 Hz.

Solution

$$Q = \frac{f_c}{F2 - F1}$$

$$Q = \frac{2000 \text{ Hz}}{(2200 \text{ Hz} - 1800 \text{ Hz})}$$

$$Q = \frac{2000 \text{ Hz}}{400 \text{ Hz}} = 5 \qquad \blacksquare$$

The filter in the above example is clearly a wideband type. It has a bandpass of 400 Hz, and a Q of 5. Now compare the Q of the narrowband filter in the next example.

EXAMPLE 9-10

Determine the Q of a filter that has a center frequency of 2000 Hz, and −3 dB frequencies of 1975 Hz and 2125 Hz.

Solution

$$Q = \frac{f_c}{F2 - F1}$$

$$Q = \frac{2000 \text{ Hz}}{(2125 \text{ Hz} - 1975 \text{ Hz})}$$

$$Q = \frac{2000 \text{ Hz}}{150 \text{ Hz}} = 13$$ ∎

The *shape factor* of the filter characterizes the slope of the roll-off curve, so is obviously related to the order of the filter. The shape factor is defined as the ratio of the −60 dB bandwidth to the −3 dB bandwidth:

$$SF = \frac{BW_{-60 \text{ dB}}}{BW_{-3 \text{ dB}}} \quad (9\text{-}44)$$

9-5.1 First-order VCVS bandpass filters

A wideband first-order bandpass filter response is obtained by cascading first-order high-pass and low-pass filter circuits, as shown in Fig. 9-17A. This arrangement overlays, or superimposes the frequency response characteristics of both filter stages. Figure 9-17B shows the situation when cascade high and low-pass filters are used. The low-pass filter response (solid line) is from DC to the −3 dB point at $F2$. The high-pass filter response is from the highest possible frequency within the range of the circuit down to the −3 dB point at $F1$. The passband is the intersection of the two sets: high and low-pass characteristics, which falls between $F1$ and $F2$.

The gain of the overall bandpass filter within the passband is the product of the two individual gains: $A_{vt} = A_{VL} \times A_{VH}$. The gain magnitude term of this form of filter is found by:

$$\left|\frac{V_o}{V_{in}}\right| = \frac{A_{vt}(f/F1)}{[1 + (f/F1)^2]\sqrt{1 + (f/F2)^2}} \quad (9\text{-}45)$$

where:
  V_o is the output signal voltage
  V_{in} is the input signal voltage
  f is the applied frequency
  $F1$ is the lower −3 dB point frequency
  $F2$ is the upper −3 dB point frequency
  A_{vt} is the total cascade gain of the filter

Cascading low and high-pass filter sections can be used to make wideband filters, but because of component tolerance and problems it becomes less useful as Q increases above about 10 or so. For narrow band filters a multiple feedback path (MFP) filter circuit such as Fig. 9-18 can be used. This filter

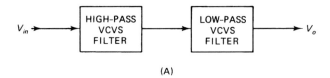

(A)

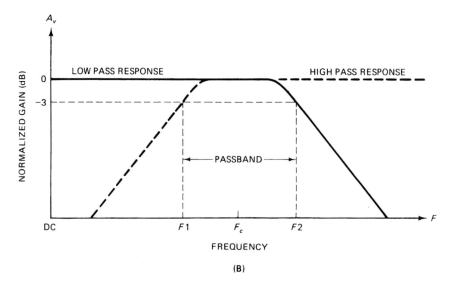

(B)

FIGURE 9-17 Combining high-pass and low-pass filters to make a bandpass filter.

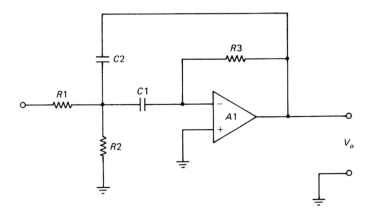

FIGURE 9-18 MFP bandpass filter circuit.

310 ACTIVE FILTER CIRCUITS

circuit offers first-order performance and relatively narrow passband. The circuit will work for values of $10 \leq Q \leq 20$, and gains up to about 15. The center frequency of the MFP bandpass filter is:

$$f_c = \frac{1}{2\pi}\sqrt{\frac{1}{R3C1C2}\left(\frac{1}{R1}+\frac{1}{R2}\right)} \qquad (9\text{-}46)$$

In order to calculate the resistor values it is necessary to first set the passband gain (A_v) and the Q. It is the general practice to select values for $C1$ and $C2$, and then calculate the required resistances for the specified values of f_c, A_v and Q. The resistor values are:

$$R1 = \frac{1}{2\pi A_v C2 f_c} \qquad (9\text{-}47)$$

$$R2 = \frac{1}{2\pi f_c(2Q^2 - A_v)C2} \qquad (9\text{-}48)$$

$$R3 = \frac{2Q}{2\pi f_c C2} \qquad (9\text{-}49)$$

and the gain:

$$A_v = \frac{R3}{R1\left[1+\dfrac{C2}{C1}\right]} \qquad (9\text{-}50)$$

These equations can be simplified if the two capacitors are made equal ($C1 = C2 = C$), and assuming that $Q > \sqrt{(A_v/2)}$:

$$R1 = \frac{Q}{2\pi f_c A_v C} \qquad (9\text{-}51)$$

$$R2 = \frac{Q}{2\pi f_c C(2Q^2 - A_v)} \qquad (9\text{-}52)$$

$$R3 = \frac{2Q}{2\pi f_c C} \qquad (9\text{-}53)$$

$$A_v = \frac{R3}{2R1} \qquad (9\text{-}54)$$

EXAMPLE 9-11

Design an MFP bandpass filter with a gain of 5 and a Q of 15 when the center frequency is 2200 Hz. Assume that $C1 = C2 = 0.01$ µF.

Solution

1. Calculate $R1$.

$$R1 = \frac{Q}{2\pi f_c A_v C}$$

$$R1 = \frac{15}{(2)(3.14)(2200 \text{ Hz})(5)(0.01 \times 10^{-6} \text{ farads})}$$

$$R1 = \frac{15}{0.00069} = 21\,714 \text{ ohms}$$

2. Calculate $R2$.

$$R2 = \frac{Q}{2\pi f_c C(2Q^2 - A_v)}$$

$$R2 = \frac{15}{(2)(3.14)(2200 \text{ Hz})(0.01 \times 10^{-6})((2)(15)^2 - (5))}$$

$$R2 = \frac{15}{0.062} = 244 \text{ ohms}$$

3. Calculate $R3$.

$$R3 = \frac{2Q}{2\pi f_c C}$$

$$R3 = \frac{(2)(15)}{(2)(3.14)(2200 \text{ Hz})(0.01 \times 10^{-6} \text{ farads})}$$

$$R3 = \frac{30}{0.000138} = 217\,140 \text{ ohms}$$ ∎

The MFP bandpass filter is capable of being tuned using only one of the resistors. If $R2$ is varied, then the center frequency will shift, but the bandwidth, Q and gain will remain constant. To scale the circuit to a new center frequency using only $R2$ as the change element select a new value of $R2$ according to:

$$R2' = R2 \left[\frac{f_c'}{f_c}\right]^2 \qquad (9\text{-}55)$$

EXAMPLE 9-12

Calculate the new value of $R2$ that will force the MFP filter in the previous example from 2200 Hz to 1275 Hz.

Solution

$$R2' = R2 \left[\frac{f_c'}{f_c}\right]^2$$

$$R2' = (240 \text{ ohms}) \left[\frac{2200 \text{ Hz}}{1275 \text{ Hz}}\right]^2$$

$$R2' = (240 \text{ ohms})(2.98) = 715 \text{ ohms}$$ ∎

9-6 BAND REJECT (NOTCH) FILTERS

A band reject or notch filter is used to pass all frequencies except a single frequency (or small band of frequencies). An application for this circuit is to remove 50/60 Hz interference from electronic instruments. The medical electrocardiograph (ECG) machine, for example, often suffers 50/60 Hz interference because the input leads are unshielded at the tips. These machines often include a switch selectable 50/60 Hz notch filter to remove the 50/60 Hz artifact that could result from interfering local electrical fields.

Figure 9-19A shows a typical active notch filter, while Fig. 9-19B shows the frequency response for the circuit. Note that the gain is constant throughout the frequency spectrum except in the immediate vicinity of f_c. The depth of the notch is infinite in theory, but in practical circuits precision matched components will offer -60 dB of suppression, while ordinary 'bench run' components (not precision) can offer -40 to -50 dB of suppression. The resonant frequency of this notch filter is found from:

$$f_c = \frac{1}{2\pi RC} \tag{9-56}$$

The gain of the circuit is unity, but the Q can be set according to the following equations:

$$Q = \frac{Q}{2R} \tag{9-57}$$

or

$$Q = \frac{C}{C_a} \tag{9-58}$$

EXAMPLE 9-13

Design a notch filter with a Q of 8 for a frequency of 60 Hz.

Solution

1. Select a trial value of $C1 = C2 = C$: 0.01 µF.
2. Calculate the value of $R1 = R2 = R$:

$$R = \frac{1}{2\pi RC}$$

$$R = \frac{1}{(2)(3.14)(60 \text{ Hz})(0.01 \times 10^{-6})}$$

$$R = \frac{1}{3.768} = 265\,392 \text{ ohms}$$

3. Select $R3$:

$$R3 = R/2$$

$$R3 = \frac{265\,392 \text{ ohms}}{2} = 132\,696 \text{ ohms}$$

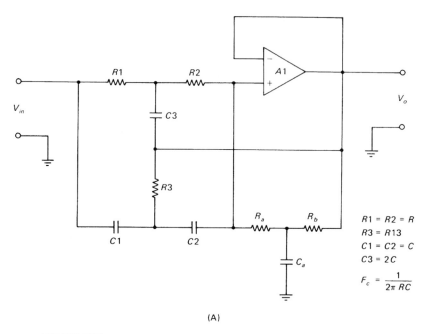

(A)

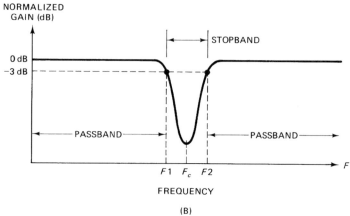

(B)

FIGURE 9-19 (A) Twin-tee notch filter circuit; (B) response.

4. Select $C3$:

$$C3 = 2C$$
$$C3 = (2)(0.01 \ \mu F) = 0.02 \ \mu F$$

5. Select R_a:

$$R_a = 2QR$$
$$R_a = (2)(8)(265\,392 \text{ ohms}) = 4.24 \text{ megohms}$$

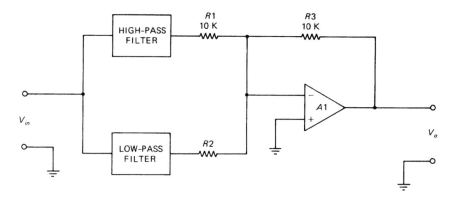

FIGURE 9-20 Bandstop filter circuit.

6. Select C_a:

$$C_a = C/Q$$
$$C_a = 0.01\ \mu F/8 = 0.0013\ \mu F$$ ∎

A bandstop filter is an example of a notch filter with a wide stopband. Just as the wide bandpass filter can be made by cascading high and low-pass filter, the wideband notch (or stopband) filter can be made by paralleling high and low-pass filter sections. Figure 9-20 shows a bandstop filter in which the outputs of a high-pass filter and a low-pass filter are summed together in a two-input unity gain inverting follower amplifier circuit. The frequency response curves of the two filter sections are superimposed to eliminate the undesired band. Make the −3 dB point of the high-pass filter equal to the upper end of the stopband, and the −3 dB point of the low-pass filter equal to the lower end of the stopband.

9-7 ALL-PASS PHASE-SHIFT FILTERS

The all-pass phase-shifter (APPS) is a special category of filter in which all frequencies (within the ability of the op-amp) are passed, but are shifted in-phase a specified amount. Figure 9-21A shows the circuit for an APPS that will exhibit phase shift between input and output of −180° to 0°. If the roles of $R1$ and $C1$ are reversed, then the phase shift will be −360° to −180°. The gain response of this circuit is:

$$A_v = \frac{1 - 2\pi f R1 C1}{1 + 2\pi f R1 C1} \qquad (9\text{-}59)$$

and the amount of phase shift:

$$\Delta\phi = -2\tan^{-1}(2\pi f R1 C1) \qquad (9\text{-}60)$$

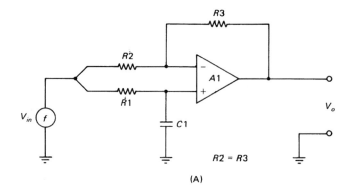

(A)

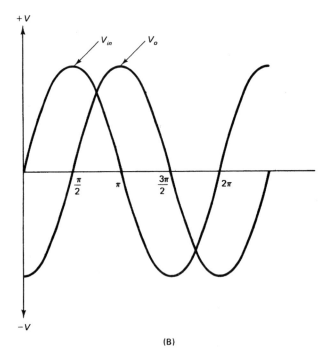

(B)

FIGURE 9-21   All-pass phase shift circuit.

9-8   STATE-VARIABLE ANALOG FILTERS

The *state-variable filter* is a variation on the multiple feedback path design using three operational amplifiers. While the circuit is more complex than the other circuits presented in this chapter, it is also more versatile. The state variable filter is capable of *simultaneously* providing bandpass, low-pass and high-pass responses. Figure 9-22 shows the block diagram of the

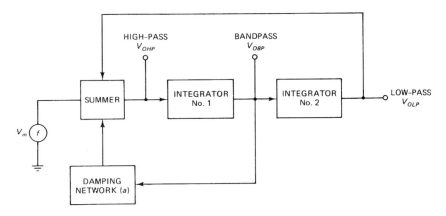

FIGURE 9-22 State variable filter in block diagram form.

state variable filter. The constituent parts include a summing amplifier, two integrators and a damping network. The integrators are inverting Miller integrators (Chapter 2). The summing amplifier is also a simple op-amp summer, and the damping network is a simple resistor voltage divider.

The damping factor (a) is the same for all three outputs, and is defined as the reciprocal of Q:

$$a = \frac{1}{Q} \tag{9-61}$$

The state variable filter shown in Fig. 9-23 is normalized for 1 kHz, $C = 0.0159\ \mu\text{F}$ and $R = 10$ kohms. Designing for other frequencies is a matter of scaling these values to the new values required for the new frequency. As in the previous examples, change either R or C, but not both. The value of

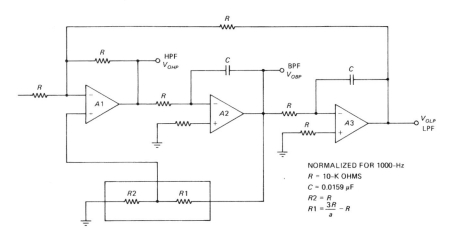

FIGURE 9-23 State variable filter circuit.

components for the damping network are found from:

$$R1 = \frac{3R}{a} - R \tag{9-62}$$

and

$$R2 = R \tag{9-63}$$

9-9 VOLTAGE-TUNABLE FILTERS

A voltage tunable filter is one in which the −3 dB response point is a function of an input control voltage. The analog multiplier and divider can be used to make a simple voltage tunable filter in either low-pass (Fig. 9-24A), high-pass (Fig. 9-24B), or bandpass (Fig. 9-24C) versions.

The low-pass filter is shown in Fig. 9-24A. This circuit consists of a Miller integrator (Chapter 2) with an analog multiplier in a second negative feedback path through R2. The X input of the multiplier is the output of the integrator, while the Y input is the control voltage (V1) that tunes the filter. The output voltage of this circuit is:

$$V_o = \frac{-V_{in}}{1 + \frac{2\pi f R1 C1}{V1}} \tag{9-64}$$

The high-pass filter circuit is shown in Fig. 9-24B. This circuit consists of an analog divider driving an RC differentiator (R1C1). The output of the differentiator is summed with the input signal. For $R2 = R3 = R$, $C1 = C$, and $R1 < R$, the following transfer equation obtains:

$$V_o = -V_{in}\left[1 + \frac{2\pi f RC}{V1}\right] \tag{9-65}$$

The bandpass circuit of Fig. 9-24C is based on the analog multiplier and a pair of integrator circuits. This circuit produces a gain of −1, and a Q of:

$$Q = \sqrt{-10\ V1} \tag{9-66}$$

The center frequency is:

$$f_c = \frac{1}{2\pi RC}\sqrt{\frac{V1}{10}} \tag{9-67}$$

(assuming a control voltage $V1 \leq 0$).
The bandwidth is:

$$BW = \frac{1}{20\pi RC} \tag{9-68}$$

All three of these multiplier or divider based circuits are dependent upon the properties of the device used. The multiplier gain, error, linearity, and response time determine the properties of the filter and the tuning rate.

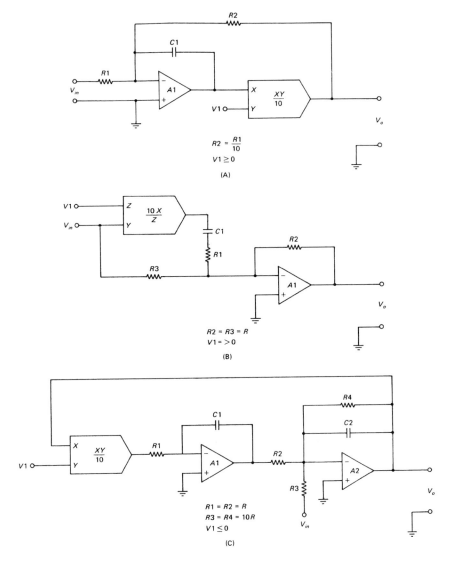

FIGURE 9-24 Voltage tunable filters.

9-10 SUMMARY

1. A filter circuit will pass some frequencies (called the passband), and either reject or severely attenuate other frequencies (called the stopband).
2. Passive filters are made from some combination of R, L, and C components. Active filter may use these same components, but also add an active device such as an amplifier.

3. Filters can be categorized according to passband: low-pass, high-pass, bandpass, stopband (or notch), and all-pass *Phase-Shifter*.
4. Another method of categorizing filters is according to passband shape. Common forms include: Butterworth, Chebyshev, Cauer (or elliptic), and Bessel.
5. The transition from passband to stopband is not abrupt, but rather falls off gradually. The slope of the roll-off is measured in either dB/octave (2:1 frequency change) or dB/decade (10:1 frequency change). Typical roll-off rates: first-order $= -20$ dB/decade, second-order $= -40$ dB/decade, and third-order $= -60$ dB/decade.
6. Filters can be designed by either the ground-up method, or the frequency scaling method. In the latter, a filter is first designed using the ground-up method for a standard frequency such as 1 Hz or 1000 Hz, and then simple ratios (f_c/f) are used to scale either the resistor or capacitor values.

9-11 RECAPITULATION

Now return to the objectives and Pre-quiz questions at the beginning of the chapter and see how well you can answer them. If you cannot answer certain questions, place a check mark to each and review the appropriate parts of the text. Next, try to answer the questions and work the problems below, using the same procedure.

9-12 STUDENT EXERCISES

In each of the following exercises you should measure the gain and the frequency response characteristics. Plot your work on graph paper, or tabulate the results if no graph paper is available.

1. Design, build and test a second-order low-pass filter circuit based on the Sallen-Key (VCVS) circuit (Fig. 9-9A). Use frequency scaling to make the -3 dB cutoff frequency 500 Hz. Measure the response for at least 25 points from 10 Hz to 10 kHz.
2. Design, build and test a high-pass filter of the same specifications as in the previous exercise.
3. Design, build and test a notch filter with a Q of 5 and a rejection frequency of 100 Hz.
4. Design, build and test a narrow bandpass filter with a Q of 15 for a center frequency of 1500 Hz.
5. Design, build and test an all-pass phase-shifter for 1000 Hz that offers a phase change of approximately 80°. How may this circuit be made variable?

9-13 QUESTIONS AND PROBLEMS

1. A filter circuit has a response that has a gain of 6 dB ± 0.25 dB from DC to 500 Hz, where the gain drops rapidly to 3 dB. The gain then rolls off to -23 dB at 5 kHz. This is a _____ -order, _____ -pass filter.

2. A low-pass filter has a −40 dB/decade roll-off from a −3 dB frequency of 1000 Hz. If the output level is 6 volts peak-to-peak at 1000 Hz, what is the output voltage at 10 kHz?

3. A _____-pass filter passes only those frequencies below the cutoff frequency.

4. A cutoff slope of −12 dB/octave represents _____ dB/decade.

5. A cutoff slope of −60 dB/decade represents _____ dB/octave.

6. A third-order Bessel filter has a phase shift maxima of _____ .

7. A frequency of 1500 Hz is applied to a first-order low-pass filter that has a −3 dB cutoff frequency (f_c) of 1000 Hz. Calculate response gain, A_{dB}, at this frequency if the mid-passband voltage gain is 15.

8. For the filter in the previous problem, calculate the gain magnitude and phase shift.

9. A first-order low-pass VCVS filter is built with the following component values: $R1 = 15$ kohms, $R2 = 2.7$ kohms and $R3 = 1.8$ kohms, $C1 = 0.015$ μF. Characterize this filter if a 1200 Hz sinewave signal is applied.

10. A low-pass filter is needed for a biomedical transducer. The cutoff frequency should be 200 Hz, and the gain should be 10.

11. Change the frequency of the normalized 1 kHz first-order low-pass filter in Fig. 9-8 to a new frequency of 200 Hz (a) by changing $R1$, and (b) by changing $C1$.

12. A second-order VCVS low-pass filter has the following component values: $R1 = R2 = 10$ kohms, and $C1 = C2 = 0.01$ μF. Calculate the −3 dB frequency (f_c).

13. Design a 500 Hz second-order low-pass filter with unity gain based on the circuit of Fig. 9-9.

14. Design a 2000 Hz second-order low-pass unity gain filter using the scaling method.

15. Design a −60 dB/decade low-pass filter for a frequency of 40 Hz. Use the scaling method, and the normalized values shown in Fig. 9-10A.

16. Design a −60 dB/decade cascade filter for a frequency of 800 Hz.

17. Calculate the cutoff frequency for a filter such as Fig. 9-14 in which $C1 = C2 = 0.0047$ μF, and $R1 = R2 = 27$ kohms.

18. Find the Q of a filter in which the center frequency is 1200 Hz, and the −3 dB points are 1250 and 1175 Hz. Is this a narrow bandpass or wide bandpass filter?

19. Find the Q of a filter in which the center frequency is 800 Hz, and the −3 dB points are 600 and 1000 Hz. Is this a narrow bandpass or wide bandpass filter?

20. Design an MFP bandpass filter with a gain of 6 and a Q of 16 when the center frequency f_c is 1000 Hz. Assume that $C1 = C2 = 0.015$ μF.

21. Scale the filter in the previous problem for a new frequency of 1500 Hz. Change only *one* component value.

22. Design a notch filter with a Q of 10 and a notch frequency of 100 Hz.

23. An all-pass phase shifter uses the following components: $R1 = 10$ kohms, and $C1 = 0.1$ μF. What is the phase shift at 1000 Hz? Express your answer in degrees.

24. Design a 100 Hz state variable filter with a Q of 3. Use the scaling method.

CHAPTER 10

Troubleshooting discrete and IC solid-state circuits

OBJECTIVES

1. Learn the types of test equipment typically needed for troubleshooting solid-state circuits.
2. Understand the difference between AC-path and DC-path troubleshooting.
3. Learn the DC method for troubleshooting circuits.
4. Learn the *signal injection* and *signal detection* methods for troubleshooting cascade circuits.

10-1 PRE-QUIZ

These questions test your prior knowledge of the material in this chapter. Try answering them before you read the chapter. Look for the answers (especially those you answered incorrectly) as you read the text. After you have finished studying the chapter try answering these questions again, and those at the end of the chapter (see Section 10-11).

1. An NPN silicon transistor common emitter amplifier is operated with a +12 Vdc potential applied through a load resistor to the collector. The voltage between the base terminal and ground is +3.4 volts DC. The 680 ohm emitter resistor is connected to ground. Calculate the emitter voltage drop. Is this transistor stage conducting? If so, what is the approximate emitter current?
2. The _____ _____ method of troubleshooting cascade circuits involves connecting a signal generator to the input, and then systematically looking for the signal at the output of each stage in succession with an oscilloscope.

3. A 0.01 Hz to 5 MHz function generator has an output impedance of 600 ohms. When troubleshooting a standard 4 MHz RF amplifier ($Z_{in} = Z_{out} = 50$ ohms) it is noted that the signal voltage is considerably lower than one would expect. The fault could be traced to improper selection of test equipment. True or false?
4. List the types of test equipment typically used in solid-state troubleshooting.

10-2 INTRODUCTION

It is one of those unfortunately facts of life that electronic circuits do not always work as expected. Whether in older equipment, that worked well for a long time before failing, or in newly constructed circuits that needs to be 'de-bugged', certain troubleshooting skills are needed. The skills involved in troubleshooting electronic circuits can be learned, and with practice become almost second nature. These skills are based on easily learned logical processes. You can, in many cases, follow step-by-step procedures to find a fault in a circuit. For sake of illustration a multi-stage cascade amplifier (Fig. 10-1) is used as a general model of how to troubleshoot.

Keep in mind that nearly all electronic circuit troubleshooting problems reduce to finding one of the following two conditions:

1. A missing, but required, path for current.
2. An added, undesired, path for current.

These current paths can be in either the AC path or the DC path of the circuit. Examples of DC path components include bias and load resistors, while the AC path components are transformers, coupling capacitors, and so forth. It is also possible for a component to be in both the AC and DC paths. Examples of the latter situation include active devices (such as transistors and ICs), diodes, and transformers that carry DC currents to the collector of a transistor.

There are several approaches to troubleshooting circuits, and we will examine them in detail below. But first, let's examine the matter of typical test equipment needed to troubleshoot solid-state circuits.

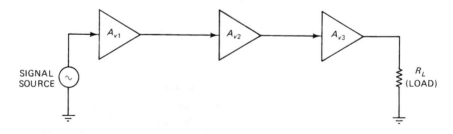

FIGURE 10-1 Generic cascade amplifier chain.

10-3 TEST EQUIPMENT

Unlike the old days of radio, when a wet finger and a good ear were considered 'test equipment', electronic troubleshooting today must be done with a variety of test instruments. The range of required instruments is extremely wide because the ranges of possible circuits and applications are wide. As a result, it is impossible in a short textbook section to lay out all of the instruments that will ever be used. It is possible, however, to make some general statements about the classes and types of instruments needed. The instruments will fall into several different classes including *signal sources*, *voltmeters* and *oscilloscopes*.

10-3.1 Signal generators

A signal generator is an instrument that produces a controllable output waveform that can be used to excite circuits for testing or troubleshooting. Although there is an immense variety of signal generators available, only a few very general categories are recognized: *audio generators*, *function generators*, and *RF generators*. While it can be argued successfully that these classes are too restrictive in the general sense, they are reasonable for our present discussion.

Audio signal generators. The audio generator produces a controllable output signal from approximately 20 Hz to 20 000 Hz. All audio generators produce a sinewave output signal, and most also produce a squarewave signal. If any particular model produces more than these two waveforms, then that model actually belongs in the function generator class.

Many commercial and broadcast audio circuits are based on a system impedance of 600 ohms. As a result, the standard audio generator is designed to have an output impedance of 600 ohms. Even if the circuit being driven is a high impedance input (e.g. as in a consumer stereo amplifier), the 600 ohm standard audio system impedance is low enough compared to the input impedance to be of negligible concern.

Two basic forms of audio generator are found, and they are different in the manner in which frequency is selected. The continuously tuned variety allows the operator to turn a single knob to select all frequencies within the set range. A separate range control is used to set the frequency range covered by the frequency setting knob. The accuracy to which this instrument can be set to a particular frequency is limited by the dial scale calibration, the width of the pointer or indicator, as well as by electronic factors such as the frequency control mechanism. Although exceptions exist, the continuously tunable generator is used on low resolution, low accuracy applications and noncritical troubleshooting. Its use in troubleshooting is limited to those applications where accuracy is either not needed or can be augmented with an external frequency counter.

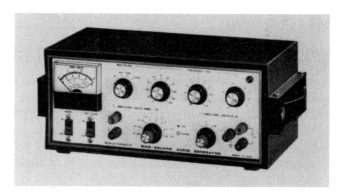

FIGURE 10-2 Step-frequency audio generator.

The other category of signal generator is the stepped frequency selection variety shown in Fig. 10-2. In this type of generator the range switch is calibrated in powers of ten (×0.1, ×1, ×10, ×100, and ×1000). There are two step-selectable sub-range switches calibrated ×10 through ×100, and ×1 through ×10 respectively. A continuously variable vernier sets the exact frequency within the range, and is calibrated 0 to 1. The set frequency is determined from the multiplication factor and adding all switch settings.

EXAMPLE 10-1

A step selectable audio generator has four controls: A, B, C and D. These controls are set as follows: A = 1000, B = 30, C = 6 and D = 0.5. Calculate the output frequency.

Solution

$$A = 1000$$
$$B = 30$$
$$C = 6$$
$$D = 0.5$$
$$E = (30 + 6 + 0.5) \times 1000$$
$$F = 36.5 \times 1000 = 36\,500 \text{ Hz}$$ ∎

The decibel scale normally used in audio work (called the *VU-scale*) measures level in *volume units* (VU), in which the reference level (0 dB or '0 VU') is defined as 1 mW at 1000 Hz dissipated in a 600 ohm load. The term volume unit is used instead of decibels to differentiate the modern scale from an archaic telephone scale that was based on a reference level definition of 0 dB as 6 mW at 1000 Hz dissipated in a 500 ohm load. The VU-scale is nonetheless a decibel scale, and the power decibels equation ($VU = 10\log[P1/P2]$) is used for VU-scale work.

One of the differences between low-cost and high-cost audio signal generators (other than the total harmonic distortion of the sinewave) is found in the output metering and control circuit. A low-cost instrument will have a poorly calibrated continuously adjustable output control, and may or may not have an output meter. Higher grade instruments use a precision attenuator with both coarse and vernier controls. The course control sets the range, while the vernier offers fine control continuously over the range. The output meter may be calibrated in either volts (rms), VU, or decibels; some instruments combine two or all three of these scales into one meter. The advantage of the precision version is that it allows exact output voltages to be produced for purposes of measuring gain or other circuit parameters.

Function generators. The function generator is much like the audio generator, except that it outputs at least sinewave, squarewave and triangle waveforms. Some function generators also output sawtooth and/or pulse waveforms. One difference between the function generator and the audio generator is that the function generator usually has a wider frequency range. While audio generators cover from 20 Hz to 20 000 Hz, with some offering wider ranges of 1 Hz to 100 kHz (as in the case of Fig. 10-2), the function generator typically offers 0.1 Hz (or less) to 2 MHz. At least one function generator is available with an 11 MHz output frequency.

Function generators generally do not have the precision attenuators or output metering that better audio generators offer. Most function generators have a 600 ohm output impedance, but some also offer in addition to the standard 600 ohm output impedance other common values: 50 ohms for RF circuits and TTL-compatible (i.e. LOW = 0 volts and HIGH > 2.4 volts) for use in digital circuits. The output metering, if used, is usually calibrated to the 600 ohm output.

The output impedance of any signal generator creates a problem in some circuits. For ideal power transfer the source impedance of the signal generator must equal the load impedance it is driving. But when several impedances are available, it is possible for an erroneous output to occur. Consider, for example, the case where a 600 ohm output signal generator is used to drive a 50 ohm load. The output meter assumes that the load is 600 ohms, so the actual output voltage will depart from the metered value by an amount determined by the voltage divider equation:

$$V_o = \frac{VR_L}{R_L + R_s} \tag{10-1}$$

Consider Fig. 10-3. The oscillator inside the signal generator produces a voltage V, which is passed to the output through source impedance R_s. The meter will read V_o accurately only if $R_L = R_s$. But when these impedances are not equal, the output meter (or setting, if no meter is used) will be in error by the amount described in Eq. (10-1).

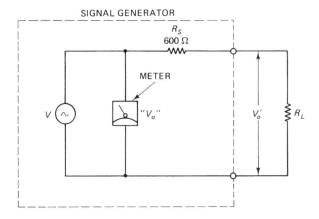

FIGURE 10-3 Signal generator output circuit.

EXAMPLE 10-2

A 600 ohm output signal generator is connected to the input of a 50 ohm circuit. Calculate the error in the output reading if this occurs.

Solution

$$V_o = \frac{R_L}{R_L + R_s} \times V$$

$$V_o = \frac{50\ \Omega}{(50\ \Omega + 600\ \Omega)} \times V$$

$$V_o = \frac{50}{650} = 0.077\ V$$

or, expressed as a percentage, the output will be 7.7% of the meter reading. ∎

RF generators. The RF generator generally produces signals above the audio range, and maybe well into the gigahertz range. Although linear IC devices only operated to about 100 MHz (and most only in the audio region) only a few years ago, there are now UHF operational amplifiers and monolithic microwave ICs (MMICs) on the market. One might, therefore, see gigahertz level RF signal generators specified to test linear IC devices. The RF signal generator may also produce signals down into the audio range. Several popular models, for example, produce signals down to 10 kHz. Very low frequency (VLF) radio and navigation stations (e.g. Omega) use these frequencies. There is no difference between a 10 kHz audio signal and a 10 kHz radio signal inside the electrical circuits. They differ in the environment, however, in that audio signals generate air pressure variations (an 'acoustic' wave), while the RF signal produces an electromagnetic (or radio) wave. The difference between an RF and an audio generator operating on the same frequency is usually the output impedance: RF generators typically produce a 50 ohm output impedance, while audio generators produce 600 ohms.

10-3.2 Multimeters and voltmeters

The multimeter (Fig. 10-4) is an instrument that will measure DC voltage, AC voltage (RMS), current and resistance. Some models will also measure certain other parameters, such as capacitance, or will offer either decibel, VU or peak reading versions of the AC voltage function in addition to the normal RMS reading. For purposes of troubleshooting, either analog or digital models are sufficient. It is important, however, to be able to measure down to tenths

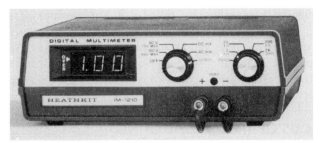

FIGURE 10-4 Digital multimeter.

FIGURE 10-5 Service grade oscilloscope.

of a volt full scale in order to correctly determine whether or nor transistors are correctly biased. Only a few applications require either extreme precision or accuracy for troubleshooting purposes (although some laboratory models require both precision and accuracy).

10-3.3 Oscilloscopes

If you are unable to have more than one instrument on a troubleshooting bench, then be sure to select an oscilloscope! The 'scope is the single most useful instrument for troubleshooting because it can measure parameters in both the DC and AC paths. Figure 10-5 shows a typical two-channel, high frequency oscilloscope designed for troubleshooting purposes. Although single-channel models are useful also, it is best to select a two-channel model because it is often necessary to compare two waveforms to inspect their time relationship.

10-4 METHODS OF TROUBLESHOOTING

There are two basic procedures for troubleshooting the circuit on a stage-by-stage basis: *signal tracing* and *signal injection*. Both of these methods involve methodically tracking a signal through successive stages in order to find where it is successfully getting through and where it is not.

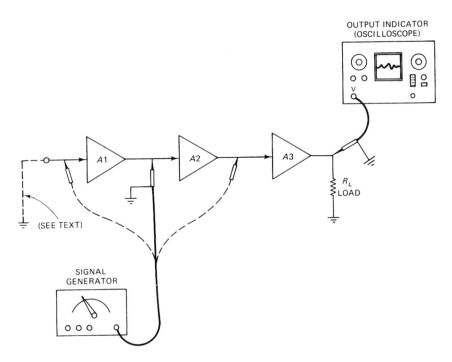

FIGURE 10-6 Signal injection troubleshooting method.

The signal injection method is shown in Fig. 10-6. In this case, the output indicator is either an oscilloscope, AC voltmeter or signal tracer across either the loudspeaker, or a 'dummy load' used in place of the loudspeaker (it's quieter that way!). The signal source is a signal generator that is capable of producing signals within the amplifier's passband. In most amplifiers or receivers, the signal generator must be capable of producing the kind of signals that the circuit responds to (sinewave, squarewave, pulse, etc.).

In signal injection, start at the output stages, and inject a signal into the input of each stage in succession. If that stage and the stages beyond it are working, then the signal will be detected on the output indicator. The signal generator output lead is then moved one stage closer to the receiver input, and the test is tried again. When the bad stage is found, it will be noted that injecting a signal either fails to produce an output indication or the output is considerably weaker than before.

The input of the amplifier may require termination when using the signal injection method. Open amplifier inputs often lead to either self-oscillation or DC latch-up of the output. In order to avoid these problems it is necessary to either short the input(s) to ground or terminate it in a specified resistance.

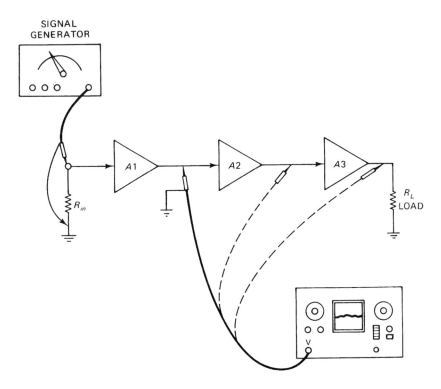

FIGURE 10-7 Signal tracing troubleshooting method.

Signal tracing technique is shown in Fig. 10-7. This method differs from the signal injection technique in that it uses a fixed signal source at the input. The detector is then moved from stage to stage to find the stage where signal is either lost or becomes very weak. Using the reverse logic of signal injection we can just as easily home in on the dead stage. One note of caution, however, signal tracers are basically high gain audio amplifiers (with a demodulator probe for detecting RF if necessary). Setting the gain control too high yields false indications when the signal is weak (due to a fault) but none the less still present.

Once the dead stage is located, other methods are used to pinpoint the particular component at fault. For example, DC methods can locate a bad transistor or linear IC.

10-5 TROUBLESHOOTING SOLID-STATE CIRCUITS

The sections above discussed methods for isolating the dead stage in a typical cascade amplifier. The methods of signal injection and signal tracing are capable of showing us which stage is bad, and from there we can do further troubleshooting to isolate the bad component. In this section we will look at a method for troubleshooting solid-state circuits. This method uses a DC voltmeter to isolate the bad stage. The method is not foolproof, but is well suited to many applications — especially when combined with other methods.

First examine Fig. 10-8. NPN and PNP transistors are shown. In most amplifier circuits the base–emitter voltage will be 0.2 to 0.3 volts for normal germanium transistors (used in older equipments), and 0.6 to 0.7 volts for silicon transistors. The following relationships also exist:

1. in a PNP transistor, the base is more negative (or less positive) than the emitter; and
2. in an NPN transistor the base is more positive (or less negative) than the emitter, depending on the polarity of the DC power supply potential.

Further, the collector will be more positive than either base or emitter in NPN transistors, and more negative in PNP transistors. Keep in mind that

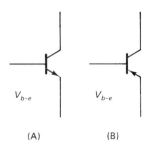

FIGURE 10-8 (A) NPN transistor; (B) PNP transistor.

the term 'more negative' can be interpreted as 'less positive' in some cases. Consider Fig. 10-9. Here we have a cascade chain of three stages. Each stage consists of a PNP transistor that is powered from a positive DC power supply. The collectors of the transistors are near ground potential, while the emitters and bases are closer to the +10.5 volt 'B+' line. If you measure the collector voltage with respect to ground you will find it very slightly positive, while the emitters are at a much higher potential. Thus, the collector being 'less positive' acts exactly like it was 'more negative.'

Voltmeter A will measure the potential between the points being examined (in Fig. 10-9 an emitter) and ground. If the voltage is near normal at each stage, then we can assume that there are no massive short circuits — but we cannot get a hint of whether the stage is working properly.

Voltmeter B is connected with its positive electrode on the B+ distribution line, and the other used to probe the emitters of the stages. The B+ line can often be located from the circuit diagram or service manual, but in some cases you will have to find it. Look for the electrolytic filter capacitor used to decouple the B+ line ($C1$ in Fig. 10-9). This filter capacitor will denote the proper line (unless you accidentally selected $C2$!), and usually has enough of a solder tab or lead wire to allow connection of the voltmeter probe.

The voltage drop across the emitter resistor of each stage indicates the current conduction of the stage. If the service manual does not give the normal voltage drop, then calculate it as the difference between the emitter potential printed on the schematic and the B+ voltage. In the case of the amplifier of Fig. 10-9, for example, the emitter voltage is 9.1 volts, while the B+ voltage is 10.5 volts. The normal conduction of this stage will be 10.5 − 9.1,

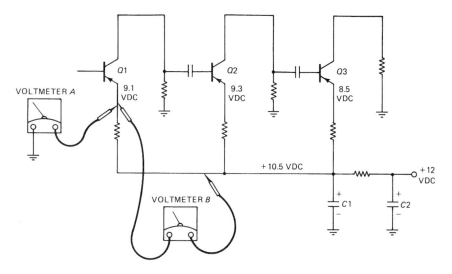

FIGURE 10-9   DC troubleshooting of PNP cascade chain.

or 1.4 volts. Any radical departure from this value indicates a problem. For example, a shorted transistor would cause that conduction voltage to increase to nearly 10 volts, while a leaky transistor would place the voltage somewhat lower but still larger than 1.4 volts. Similarly, an open emitter (or other condition that cuts off the stage) will reduce the voltage across the emitter resistor to either zero or nearly zero.

It is possible isolate the defective stage by looking at each emitter voltage in its turn. Most often, the defective stage will show up from an anomaly in the emitter conduction voltages.

Circuits with NPN transistors in the stages are similarly treated. Figure 10-10 shows a circuit with NPN transistors powered by a positive-to-ground DC power supply. The collectors of these transistors will be close to the B+ potential, while the emitter and base voltages will be a lot lower. The values shown in Fig. 10-10 are typical, but they are not to be considered absolute (there are a lot of design choices that could alter the values, so buy and consult the service manual for the particular equipment).

As in the case of the PNP transistor stages, the NPN emitter conduction voltage denotes the stage activity. In this type of circuit, however, the reference point for voltage measurement is the chassis ground instead of the B+ line. The principle is the same, however. Check each conduction voltage in its turn and determine if any of them are incorrect. Fortunately, there are fewer calculations to make in this type of circuit. The emitter voltage on the schematic is the conduction voltage.

Variable frequency oscillators will behave a little differently than amplifiers. In most cases, the DC conduction of the oscillator transistor varies with frequency setting. Typically, the voltage will be higher at the low end of the range, and lower on the high end of the range (with a smooth transition as the

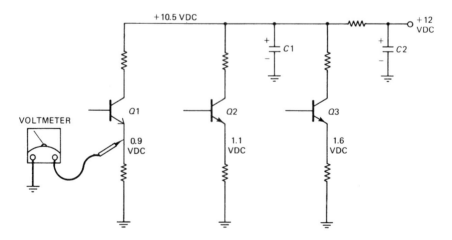

FIGURE 10-10   DC troubleshooting of NPN cascade chain.

frequency is varied). A sudden discontinuity in this transition might indicate a sudden cessation of oscillation (or a parasitic developing) at that point on the dial.

10-6 A SIMPLE SEMICONDUCTOR TESTER

With a little knowledge, the analog or digital multimeter becomes a transistor and diode tester. If you have an ohms scale, then you can measure the diode's front-to-back resistance, and from that information determine whether or not the diode is leaking. You'll even be able to tell which end of an unmarked diode is the cathode or anode.

10-6.1 Ohmmeter orientation

To understand our test method, let's first look at the test equipment. The question is 'how do analog and digital ohmmeters work?' In analog ohmmeters a battery or electronic DC power source is connected in series with the meter movement and some calibration resistors (Fig. 10-11A). Battery current pushes its way through the meter movement into an external circuit, and back to the battery. If we make it harder for the current to flow by placing a resistance in series, the meter pointer won't swing as far along the dial as when the probes are simply shorted together. That's why you'll find high precision resistors in any decent multimeter — they limit the meter pointer travel (and the current) so that the pointer always comes to rest at a specific point on the dial. Short an ohmmeter's leads together, and that place will be (of course) zero ohms.

When we connect our multimeter across an external resistance (with the multimeter in the ohms mode) the meter receives even less current than it did before; so there's less swing on the pointer's part by an amount proportional to the resistance across the terminals.

Digital multimeters (DMM) use a different tactic for measuring resistance. In Fig. 10-11B we see that a *constant current source* (CCS) is used to provide a precision reference current to the unknown resistance, R_x. The main circuit of the DMM is basically a millivoltmeter with a range of 0 to 1999 millivolts. By passing a constant current through the resistance, we can measure the resistance by measuring the voltage drop. For example, a 1 mA constant current produces a voltage drop of 1 mV/ohm. Typically, the constant current will be different for each range but the principle is the same.

10-6.2 Testing diodes

Diodes are the easiest semiconductor devices to test, so let's examine them first. Besides, this method of transistor testing depends upon the fact that the diode also represents the base–emitter and base–collector junctions of a transistor. We rely on diodes to pass electrical current *unidirectionally*. This unique ability forms the basis for our test. There are four states of being for

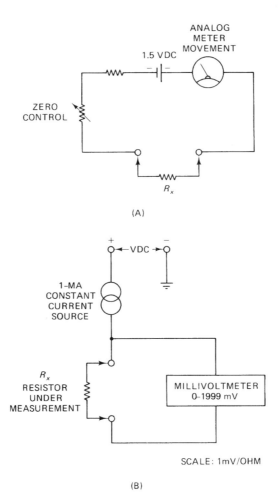

FIGURE 10-11 (A) Simple ohmmeter; (B) constant current ohmmeter.

any diode: they can be *open, shorted, leaky* or *OK*. Before you start testing, it would be wise if you had a sheet of paper and a pencil or pen to jot down the readings.

Let's first assume that we are testing a rectifier diode. With the ohmmeter set to the RX1 scale (Fig. 10-12A), place the probes across the diode leads. Record your readings, and then swap probes (Fig. 10-12B) and take a second resistance reading. After you take both readings you're ready to interpret what you've discovered.

Let's assume for the sake of simplicity that the first diode under test checked out OK. How did you know? If the diode is healthy, then one of your resistance readings will be very much higher than the other. The

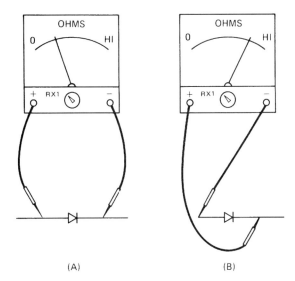

FIGURE 10-12 Using an ohmmeter to check PN diode.

actual resistance readings are not terribly important; it's the *ratio between the readings* that counts. Consider five-to-one (>5:1) or higher 'normal' for older rectifier diodes, and ten-to-one (>10:1) for newer rectifier diodes and small-signal diodes. For example, if the low reading is 500 ohms (a typical value), then the second reading should be 5000 ohms or more on a good diode. Diodes of very recent manufacture may read a very high ratio, or even infinite.

Small-signal diodes are tested in exactly the same manner as rectifier diodes, except that the meter is set to the 'RX100' scale rather than RX1. The lower RX1 scale produces a higher current flow in the external circuit, and can thus blow some small-signal diodes.

Dead-shorted diodes often have a zero ohms resistance regardless of probe polarity (Fig. 10-13). Suppose, however, that your first and second readings are almost identical (but not zero ohms). That diode is leaky, and is almost as useless.

Shorted and open diodes test exactly as you might expect. Open diodes show a very high (or sometimes infinite) resistance, as in Fig. 10-14. On analog meters the pointer doesn't budge off the 'infinity' symbol, while on digital meters the display will flash '1999'. A shorted diode will show either zero ohms in both directions, or a very low resistance (e.g. less than 100 ohms). Note that it is not always true that a shorted diode exhibits the same low resistance in both directions. In some cases, the resistances might be 50 and 80 ohms — that's very leaky and therefore to be considered 'dead shorted' for all practical purposes.

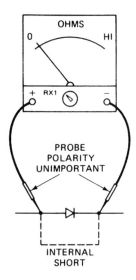

FIGURE 10-13   Reading when diode is shorted: low resistance.

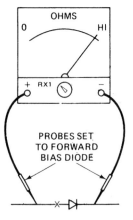

FIGURE 10-14   Reading when diode is open: high resistance.

The ohmmeter can also tell us which end of an unmarked diode is the anode or cathode. Here's where you must know the relative polarity of your ohmmeter probes. One way of determining polarity is to measure the voltage across the probes with a DC voltmeter, and note the polarity of the reading. Another method is to take a known good diode that is marked as to cathode and anode (the cathode end is usually marked somehow: stripe of paint, bullet shape, etc.) and connect it to the meter in a direction that produces a low resistance. In that state, the ohmmeter positive lead is applied to the diode anode — and could be marked. The red ohmmeter lead is usually positive.

Once you know which probe of the ohmmeter is positive, you can use that information to tag any diode as to anode and cathode. Connect the diode across the leads to produce the low DC resistance reading — the positive lead is always connected to the anode end (and, of course, the other end is the cathode).

10-6.3 Testing transistors

Transistor types can be (and often are) categorized in terms of family behavior. Look in any transistor catalog and you'll see that many consecutively numbered transistor types share like characteristics. You will find that NPN and PNP types that are identical in performance, but have equal and opposite polarities. These are called 'complementary pairs'.

Let's start by examining a small-signal PNP unit first. The method for checking transistors with VOM, VTVM, TVM or DMM is no different from that for checking a diode. Perform your tests with the ohmmeter on the RX100 scale for small transistors and the RX1 scale for power transistors. Connect your negative ohmmeter probe to the transistor's base lead. Separately touching the collector and then emitter leads with the positive lead you'll detect either a high resistance or a low resistance on each junction. Reversing the probe connections — placing the positive probe on the base — will show up the opposite reading (see Figs 10-15A through 10-15D).

By now you're wondering what these resistance readings have to do with transistor testing. First, you found out if the transistor was leaky, shorted or open between elements. This would become apparent as you made your resistance measurements and found very undiode-like behavior across the junctions. Next you either confirmed or discovered the transistor's polarity, i.e. whether it was PNP or NPN.

Suppose the base-emitter junction tests open. This situation is revealed as a very high resistance, whether the positive probe is connected to the base, or the other way around. If the base-emitter junction tests shorted, then you'll see that this condition shows up as a very low resistance, no matter which way the ohmmeter probes are placed. Leaky transistors give you a real run for your measurement money. Indicating abnormal base-emitter resistance values, they won't always test either open or shorted — normal resistance characteristics for that particular transistor family under examination simply won't show.

Finding out whether a transistor is PNP or NPN merely amounts to noting lowest resistance values as you check base-emitter junctions (refer again to Fig. 10-15). For PNP units, the lowest reading occurs when the positive probe is connected to the emitter, and the negative lead is hooked to the base. For NPN transistors it is just the opposite. That is, you'll get a low reading when the positive lead is connected to the base and the negative is connected to the emitter.

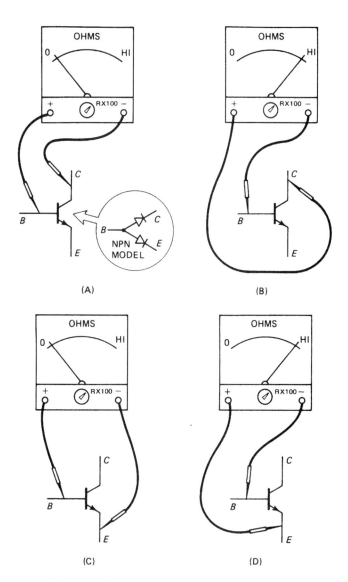

FIGURE 10-15 Testing a transistor using an ohmmeter to check the c–e and b–e junctions.

10-6.4 Testing for `gain´

No transistor is worth much if its base terminal does not control collector–emitter current flow. All of the test methods listed above will give us some gross indications of failure — an open emitter, shorted C–E junction and so forth — but don't really tell us anything about whether or not the transistor will amplify. Figure 10-16 shows a method used by some technicians to determine whether or not a transistor is good. Connect

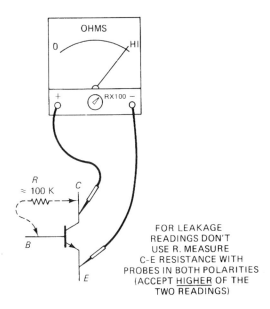

FIGURE 10-16 Using an ohmmeter to test c–e leakage.

the ohmmeter across the collector–emitter leads in the polarity that would normally be found in circuit. For PNP units connect the negative probe to the collector, and the positive probe to the emitter; for NPN units connect the positive probe to the collector and the negative probe to the emitter. If the transistor is good, the reading will be very high, or infinity. Next, connect the base lead to the collector lead through a 100 kohm resistor. If the transistor works, then the resistance reading across C–E terminals will drop.

10-6.5 Checking leakage

Leakage in a transistor is the unwanted flow of current from collector to emitter (or vice versa). Checking a transistor for leakage is simply measuring resistance across C–E twice. Measure first with one polarity, and then switch the ohmmeter leads and measure again. The leakage reading is the higher of the two resistances (because of junction action, one reading is normally quite low). For germanium transistors the leakage will be lower than for silicon. Typically, silicon transistors will show nearly infinite resistance on the three scales normally used: RX1, RX10 and RX100. Germanium transistors may have 100 k of leakage and still be good enough for use.

10-6.6 Using digital multimeters

The digital multimeter (DMM) has largely (but not entirely) replaced the old-fashioned VOM/VTVM. Earlier we discussed the method used in DMMs to measure resistance. Unfortunately, that method does not lend itself well to our semiconductor test method because the voltage produced across the probe tips is not sufficient to forward bias PN semiconductor junctions. Although this feature makes it easy to make in-circuit resistance measurements without removing the semiconductors, it also prevents us from testing semiconductors. However, most DMMs are designed to overcome that problem. There will be a special ohms scale that will forward bias diode junctions. These scales are marked with either words such as 'High Power', or (more commonly) with the diode symbol.

10-6.7 Safety rule

Diodes and small transistors can be damaged by ohmmeter currents. Always start at the higher scales ($\geq$RX100) and then drop down to lower scales (RX10 or RX1) only if necessary to get a readable deflection of the meter. In any event, stay on the same scale for both readings. The typical ohmmeter circuit changes currents on different ranges, so comparing readings taken on different readings means interpreting the results of two different bias levels — and that's comparing apples and mangos. Also, when using an old VTVM, e.g. one made before about 1965, then be certain that the ohmmeter battery is 1.5 Vdc, not 22.5 Vdc (which will blow out most semiconductors).

10-7 TESTING LINEAR IC CIRCUITS

Linear IC devices can be both easier and harder to troubleshoot than ordinary transistor circuits, depending upon the situation. They are sometimes harder to troubleshoot because they are seen as block functions in the schematic, so you may not have the insight into internal circuit workings as is normal with discrete transistor circuits. This is one good reason to keep the data books and specifications sheets for linear devices nearby when troubleshooting. At the same time, however, the very fact that the device is a block makes it easier to rapidly determine whether or not it is faulty.

As stated earlier, there are two paths in any circuit: AC (or signal) and DC. In the sample circuit of Fig. 10-17 the capacitors ($C1$ through $C4$) and the operational amplifier ($A1$) are in the AC path for current. Capacitors $C1$ and $C2$ are coupling capacitors, while $C3$ and $C4$ are decoupling or bypass capacitors. The DC path consists of the DC power supply terminals (pins 4 and 7), the input pins (2 and 3) and the output pin (6). Let's examine the troubleshooting logic under several different scenarios.

Condition 1. The operational amplifier will not pass signal when a 1000 Hz AC signal at 1 volt p–p is applied to the input. If an oscilloscope is available, examine the input (point A) to see if, indeed, the signal generator is supplying signal (many a headache is caused by defective test equipment). Next, move the probe to point B and make sure the signal is passing through

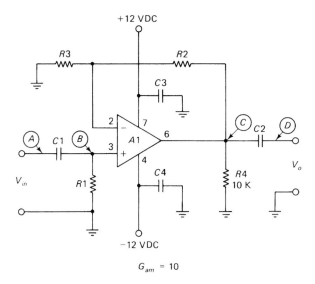

FIGURE 10-17 Audio amplifier circuit.

input coupling capacitor C1. If the signal is passing through C1, then move the 'scope probe to the output (point C) and check for signal. If signal appears at point C, but not at the output (point D), then the output coupling capacitor is probably open. Check by substitution.

Condition 2. The same scenario as Condition 1, but in this case it is found that there is signal at points A and B, but not at point C. This situation suggests that the operational amplifier is not passing the signal. Now it is time to examine the various operational amplifier terminals for DC levels. Check the $V-$ and $V+$ power supplies (pins 4 and 7, respectively). The voltage should be within tolerance (usually $\pm 15\%$ if unregulated, or less if regulated).

Assuming that the DC supplies are present, next check the output terminal (pin no. 6). Assume that a high DC voltage close to either $V-$ or $V+$ is found. It is possible to determine whether or not the operational amplifier is working by determining whether or not the inputs are capable of controlling the output. Short $+$IN and $-$IN together while monitoring the output pin. If the op-amp is good, then the voltage will drop to zero when pins 2 and 3 are shorted together. If the voltage does not change, then assume a fault and replace the IC.

Condition 3. No signal passes, and output pin no. 6 is found to be at a potential of $+10$ volts. Measurement of the power supply pins reveals that pin no. 7 is at $+12$ Vdc, while pin no. 4 is at ground potential. The problem here is loss of the $V-$ DC power supply.

Condition 4. A high frequency oscillation is found riding on the output waveform. The DC voltages are found to be normal, or very nearly so. To find the cause of this problem shunt known-good capacitors across $C3$ and $C4$ while monitoring the output on an oscilloscope. If the oscillation ceases when a good capacitor is shunted across the old capacitors, then one or both capacitors are probably bad. It must be noted, however, that other problems can cause oscillation, especially in circuits with feedback networks more complex than this one.

10-8 SUMMARY

1. Most circuit troubleshooting involves finding either (a) a lost path for current, or (b) an undesired path for current.
2. Circuits contain both AC and DC paths for currents, and the troubleshooter must analyze both.
3. Cascade amplifiers can be approached using either signal injection or signal tracing.
4. An ohmmeter can be used to test diodes and transistors, although it is less useful for linear IC devices.

10-9 RECAPITULATION

Now return to the objectives and Pre-quiz questions at the beginning of the chapter and see how well you can answer them. If you cannot answer certain questions, place a check mark to each and review the appropriate parts of the text. Next, try to answer the questions and work the problems below, using the same procedure.

10-10 QUESTIONS AND PROBLEMS

1. List two different approaches to troubleshooting a cascade amplifier.
2. Can an ordinary digital ohmmeter be used to check a PN junction diode?
3. You can afford only one test instrument on your workbench. Select from the following: signal generator, multimeter or oscilloscope.
4. A transistor amplifier uses a PNP transistor in a circuit with a negative ground DC power supply. The collector DC path returns to ground through a 1000 ohm resistor. The 680 ohm emitter resistor is connected to a $+11$ Vdc power supply line. The voltage from emitter to ground is 10.2 Vdc. Calculate: emitter resistor voltage drop, emitter current, and approximate DC voltage that appears from collector to ground.

Index

A/D, 33, 82, 148, 163.
Absolute value circuit, 13, 266
AC amplifier, 130
AC amplifier, narrowband, 35
AC signals, 4
AC-coupled trigger (555 timer), 124
AC-excited amplifier, 37
AC-excited carrier amplifier, 41
Accuracy, 142
Active filter circuits, 281.
Active integrator/differentiator circuits, 47
ADC, see A/D
AGC, 276
All-pass filter, 315
AM demodulation, 9
AM demodulator, 9, 238
AM modulator, 228
Amplifier classes, 199
Amplifier drift cancellation, 40
Amplifier, biopotentials, 33
Amplifier, bridge, 206
Amplifier, carrier, 37
Amplifier, chopper, 34
Amplifier, control, 176, 192
Amplifier, laboratory, 33
Amplifier, limiting, 236
Amplifier, lock-in, 40
Amplifier, power, 176, 199
Amplifier, programmable gain, 60
Amplifier, transducer, 84
Amplitude modulation, 38, 211, 224
Amplitude modulator, 39
AMV, 114
AMV (555 timer), 133
Analog multiplier, 60, 454
Analog reference, 154
Analog to digital comparator, see A/D
Anti-log amplifier, 57
Anti-log circuit, 52
ASCII, 212, 247
Astable multivibrator, see AMV
Asymmetrical bipolar coding, 152
Asynchronous demodulation, 228
Audio, 175
Audio circuits, 600-ohm, 185
Audio generator, 325

Audio mixer, 175, 187
Auto-zero circuit, 76
Autocorrelation amplifier, 41
Automatic gain control (AGC), 276
Avalanche, 28

Balanced modulator, 229, 231
Band reject filter, 313
Bandpass, 40
Bandpass filter, 238, 307
BAUDOT, 212
Baxandall tone control, 197
Bessel filter, 286
BiFET, 24
BiMOS, 24
Binary coded decimal (BCD code), 140
Binary counter, 31, 167
Biopotentials amplifier, 33
Bipolar coding, 152
Blood O_2 level, 103
Blood saturation, 103
Bonded strain gage, 69
Bounded amplifier, 2
Breakpoint generator, 80
Bridge, 67
Bridge amplifier, 206
Bridge power amplifier, 175
Bridge transducer, calibrating, 75
Buffer, 22
Butterworth filter, 285

Calibrating bridge transducer, 75
Calibration resistor, 76
Camera flashgun, 127
Capture mode (PLL), 240
Cardiotachometer, 129
Carrier amplifier, 37
Carrier amplifier, AC-excited, 41
Carrier frequency, 38
Carrier signal, 9, 39
Cauer filter, 285
CDA mixer, 190

Chebyshev filter, 285
Chopper amplifier, 34
Chopper amplifier, differential, 35
Chopper frequency, 37
Chopper mechanical, 35
Clamp circuit, 25
Clamper, 2
Class-A amplifier, 199
Class-AB amplifier, 199
Class-B amplifier, 199
Clipper, 2
Clipper circuit, 25
Clipping, 238
Clipping action, 18
CMOS electronic switch, 25, 51, 61
CMOS preamplifier, 184
CMRR, 89
Coding, data converter schemes, 152
Colorimeter, 100
Colorimetry, 95, 102
Common mode rejection ratio, see CMRR
Communications, 211
Comparator, 163
Compensation, empty weight, 76
Compensation, temperature, 110
Complementary symmetry metal oxide semiconductor, see CMOS
Complementary symmetry push-pull power amplifier, 201
Compression amplifier, 191
Constant current source, see CCS
Control amplifier, 176, 192
Control system, feedback, 240
Conversion gain, 242
Counter-type A/D, 168
CRO, 111, 330
Crossover distortion, 201
Crystal oscillator, 238
Current difference amplifier, see CDA
Current loop (4-to-20 mA), 215
Current loop, 211

Current loop serial communications, 212
Current pump, 218
Current, leakage, 24
Cuvette, 104

DAC, 60, 77, 148
DAC, IC example, 153
DAC, multiplying (MDAC), 151
Dark resistance, 100
Data coding schemes for DAC and A/D, 152
Data converter circuits (A/D and DAC), 148
Data logging, 33
DB-25 connector, 222
DC amplifier, 239
DC differential amplifier, 16, 90, 156, 247
DC differential circuit, 187
DC feedback control, 248
DC level, 236
DC power supply, 14, 27
DC-excited amplifier, 37
Dead-band amplifier, 21
Dead-band circuit, 16, 20
Demodulation, 9, 224
Demodulator, AM, 238
Demodulator, FM, 236
Densitometer, 100
Derivative, 44
Detection, envelope, 9
Detection, quadrature phase, 244
Detector, envelope, 38
Detector, synchronous, 238
Diastolic, 43
Differential amplifier, DC, 16, 90, 156, 247
Differential chopper amplifier, 35
Differential input multiplier/divider, 259
Differentiation, 44, 51
Differentiator, 41, 51
Differentiator, active, 47

Digital phase detector, 239
Digital-to-analog converter, see DAC
Diode applications, 1
Diode breakpoint generator, 80
Discriminator, polarity, 11
Distortion, crossover, 201
Divider, analog, 254
Divider, differential input, 259
Double sideband suppressed carrier (DSBSC), 231
Drift cancellation, 40
Dual conversion, 236
Duty cycle (555 timer), 136
Duty-cycle modulator, 234
Dynode, 97

E-field noise, 40
ECG, 131
Einstein, Albert, 97
Electrocardiograph, see ECG
Electrode, glass, 107
Electrodes, 68
Electromagnetic radiation, 95
Electrometer amplifier, 109
Electronic instrumentation, 33
Electronic switch, 25, 51, 61
Elliptic filter, 285
Empty weight compensation, 76
End-of-conversion (EOC), 165
End-of-conversion, 170
Envelope detection, 9, 228
Envelope detector, 38
Equation, transfer, 28, 52
Excitation signal, 18
Excitation source, 82
Excitation voltage, 74
Exponentiator, voltage controlled, 277

Feedback control system, 240
Feedback control, DC, 248
Feedback loop,

Feedback loop, 22, 28
Feedback network, 78
Feedback network, treble, 195
Feedback path, 11
Feedback resistor,
Feedback resistor, 27
Filter, 41, 46
Filter characteristics, 283
Filter phase response, 286
Filter, active, 281
Filter, all-pass, 315
Filter, band reject, 313
Filter, bandpass, 238, 307
Filter, high-pass, 303
Filter, loop, 248
Filter, low-pass, 238, 244
Filter, low-pass, 287
Filter, notch, 313
Filter, phase shift, 315
Filter, state variable, 316
Filter, voltage tunable, 318
Flame photometer, 104
Flash A/D, 17
Floating load, 145
FM demodulator, 236
FM detector, 131
Free-running mode (PLL), 240
Frequency meter, sub-audio, 249
Frequency modulation, 224
Frequency of oscillation, 135
Frequency response, 178
Frequency shift keying, 238
Fullwave precise rectifier, 13
Function generator, 327

Gain control, master, 189
Gain, conversion, 242
Galena crystal, 2
Generator, function, 327
Generator, signal, 325
Germanium, 2
Glass electrode, 107
Graphic equalizer, 194
Ground loop noise, 40

Ground referenced load, 145

H-field noise, 40
Half-bridge circuit, 70, 100
Harmonic distortion, 204
Harmonics, 38
Hash, 35
High-pass filter, 41, 46, 303
Howland current pump, 218
Hybrid, 37

Ice bath, 63
Ice point compensation, 63
in vitro, 89
Infrared, 95
Instantaneous rate of change, 44
Instrumentation circuits, 67
Instrumentation, electronic, 33
Integration (def.), 42
Integration A/D, 163
Integrator, 32, 40, 41, 48, 159
Integrator, active, 47
Integrator, practical circuits for, 50
Integrator, RC, 46

Junction potential, 4

KCL, 4, 16, 47, 57, 219
Kelvin (temp.), 56
Kirchhoff's current law, see KCL

Laboratory amplifier, 33
Lamp multiplier, 260
Leakage current, 24
Light transducer, 95
Light/dark resistance ratio, 100
Lightmeter, 100
Limiting amplifier, 236
Line driver amplifier, 185
Line receiver amplifier, 185
Line transformer (600-ohm), 185

Linearization, 80, 162
Linearization, transducer, 78
Liquid level detector, 126
Lock-in amplifier, 40
Locked mode (PLL), 240
Logarithmic amplifier, 32, 52
Logarithmic amplifier, 263
Loop filter circuits, 248
Loop, feedback, 22, 28
Low-pass filter, 41, 46, 233, 236, 244, 287
Low-pass filtering, 9
Lower sideband (LSB), 227

Manometer, 76
Master gain control, 189
Matched resistor multiplier, 260
Mean arterial pressure (MAP), 44
Measurement circuits, 67
Mechanical vibrator, 37
Metallic tip contact, 91
Microelectrode, 91
Microelectrode capacitance, neutralizing, 94
Microphone preamplifier, 177
Miller integrator, *see* integrator
Missing pulse detector, 127
MIT Lincoln Laboratories, 218
Mixer, audio, 175, 187
MMIC, 328
MMV, 114, 127
Modulation, 224
Modulation input pin, 140
Modulator, AM, 228
Modulator, duty-cycle, 234
Monostable multivibrator, *see* MMV
Multi-function converter, 58
Multiplier circuits, types of, 260
Multiplier, analog, 60
Multiplier, analog, 254
Multiplier, circuit symbols, 255
Multiplier, differential input, 259
Multiplier, IC example, 269

Multiplier, logarithmic amplifier based, 263

NAB tape preamplifier, 182
Narrowband AC amplifier, 35
Noise, 34, 40, 238
Nonlinear applications, 1
Norton amplifier, *see* CDA
Notch filter, 313
Null potential, 40

Offset control, 109
One-shot, *see* MMV
Operational amplifier, *see* op-amp
Operational transconductance amplifier, *see* OTA
Optocoupler, 192
Optoisolator, 222
Oscillation, 54
Oscillation, frequency of, 135
Oscillator, crystal, 238
Oscillator, pendulum, 137
Oscilloscope, 330
Oven, temperature stabilization, 240
Overvoltage protection, 82

PAM, 224
Parallel A/D, 171
Peak follower, 22
Peltier effect, 63
Pendulum oscillator, 137
pH electrode, 107
pH meter, 107
Phase detector, 238, 274
Phase detector, digital, 239
Phase locked loop (PLL), 238
Phase locked loop, IC devices, 244
Phase meter, 274
Phase modulation, 224
Phase response, filter, 286
Phase sensitive detector, *see* PSD
Phase shift filter, 315

Phonograph, 176, 182
Photo detector, 104
Photocell, 103
Photocolorimetry, 102
Photoconductive cell, 98
Photodiode, 101
Photoelectric effect, 97
Photoemission, 97
Photographic light meter, 100
Photography, wildlife, 127
Photometer, flame, 104
Photomultiplier tube, 97
Photon, 96
Photoresistor, 98, 100
Photosensor, 96
Photospectrometer, 104
Phototransistor, 101
Phototube, 97
Photovoltaic cell, 98
Piezoresistive strain gage, 68
Piezoresistivity, 68
PIN diode, 37
Planck, Max, 96
PLL modes, 240
Plus zero, 153
PN junction diode, 2
Pneumotachometer, 129
Polarity discriminator, 11
Power amplifier, 176
Power amplifier, IC, 204
Power amplifiers, 199
Power supply rejection ratio, see PSRR
PPM, 224
Preamplifier, 176
Preamplifier, 177
Preamplifier, audio example (LM-381), 181
Preamplifier, CMOS, 184
Preamplifier, NAB tape, 182
Precise diode circuitry, 4
Precise rectifier, 2, 18, 22, 266
Precise rectifier, fullwave, 13

Precision voltage reference (DAC), 149
Pressure transducer, 76
Process control, 215
Process instrumentation, 212
Product detector, 231, 233
Programmable gain amplifier, 60
PSD, 38, 241, 248
Pulse amplitude modulation, 224
Pulse counting detector, 236
Pulse counting FM detector, 131
Pulse position circuit, 129
Pulse repetition rate, 131
Pulse width modulation, 224
Pulse width modulator circuits, 234
Pulse, missing detector, 127
Push-pull amplifier, 201
PWM, 224, 234

Q, 40, 308
Quadrature detector, 236, 238
Quadrature phase detection, 244
Quanta, 96
Quantum mechanics, 96
Quarter square multiplier, 264
Quasi-complementary power amplifier, 202

R-2R ladder network, 150
Ramp A/D, 168
Ramp generator, 163
Rate-of-change circuit, 41
Ratiometric colorimeter, 106
Ratiometric measurements, 105
RC components, 91
RC integrator, 131
RC integrator circuits, 46
RC low-pass filter, 233
RC network, 191
RC time constant, 48, 122
RC timing network, 145
Read only memory (ROM), 162
Receivers, radio, 236
Reference electrode, 91

Reference frequency, 238, 240
Reference voltage, 152
Reference, analog, 154
Resistance, dark, 100
Respiratory CO_2 level, 104
Retriggerable MMV operation (555 timer), 126
RF generator, 328
RIAA, 182
RS-232C, 211, 221

S/H, 25
Sample and hold circuit, *see* S/H
Sawtooth generator, 137
Sawtooth generator, digital synthesized, 159
Scale factor, 85
Schmitt trigger, 130, 168, 236
Seebeck effect, 61
Self-heating, 82
Sensitivity factor, 74, 76
Sensitivity, transducer, 74
Serial voltage level communications, 212
Servo-type A/D, 168
Servomechanism, 240
Shelf equalizer, 193
Signal generators, 325
Signal injection, 330
Signal processing circuits, 32
Signal tracing, 330
Signal-to-noise ratio, *see* SNR
Single conversion, 236
Single sideband suppressed carrier, 231
Small-signal diode, 228
SNR, 89
Special function circuits, 58
Spectrometer, 104
Square rooter, 275
Square wave generator (555 timer), 133
Stability, 142
Staircased voltage ramp, 161

Standing wave ratio, *see* SWR
Start command, 165
State variable filter, 316
Strain gage, 67, 69
Strain gage circuitry, 71
Successive approximation A/D, 169
Superheterodyne receiver, 236
Surface mount package, *see* SMD
Synchronization, XR-2240 timer, 141
Synchronized oscillator, 136
Synchronous demodulation, 228
Synchronous detector, 238
Systolic, 43

Tachometry, 129
Teletypewriter, 212
Temperature, 56
Temperature compensate, 56
Temperature compensation, 110
Temperature sensitivity, 57
Temperature sensor, 61
Temperature stabilization oven, 240
Test equipment, 325
THD, 204
Thermal drift, 34, 82
Thermocouple, 61
Thermocouple amplifier, 61
Thermocouple letter designations, 63
Thermodynamics, 96
Thevenin's looking back resistance, 90
Time averager circuit, 41
Time constant, RC, 48, 122
Timebase, 142
Timebase circuit, 138
Timer, IC, 114
Timer, very long duration, 141
Timers, other IC, 144
Timing equation (555 timer), 122
Tone control circuit, 192
Tone control, reactance, 197
Total harmonic distortion, 204

Touchplate trigger, 124
Touchtone, 238
Transconductance cell, 267
Transconductance multiplier, 266
Transducer, 39, 68, 73
Transducer amplifier, 84
Transducer excitation source, 82
Transducer linearization, 78
Transducer resistance elements, 40
Transducer sensitivity, 74
Transducer, light, 95
Transducer, pressure, 76
Transfer characteristic, 2
Transfer equation, 28, 52
Transistor-transistor-logic, see TTL
Treble roll-off, 193
Triangle waveform, 162
Trigger, touchplate, 124
Triggering (555 timer), 122
Triggering, 138
Troubleshooting methods, 330
Troubleshooting, 323

Unbonded strain gage, 69
Upper sideband (USB), 227

V/F converter, 167

Varactor, 37
Variable frequency oscillator, 239
VCO, 159, 167, 239, 245
VFO, 239
Vibrator, mechanical, 37
Voltage comparator, 120, 236
Voltage controlled exponentiator, 277
Voltage controlled oscillator, see VCO
Voltage divider, 72
Voltage ramp, 163
Voltage regulation, 82
Voltage regulator, 84
Voltage tunable filters, 318
Voltage-to-frequency converters, see V/F
VSWR, see SWR

1-LSB value, 151, 153
1-LSB voltage, 161
4-to-20 mA current loop, 215
555 AMV, 133
555 MMV operation, 119
555 timer, 115
555 triggering methods, 122
600-ohm audio circuits, 185